Floods, Famines,
and Emperors

Also by Brian Fagan

Floods, Famines, and Emperors

Emperors

El Niño and the Fate of Civilizations

Brian Fagan

BASIC
BOOKS

A Member of the Perseus Books Group

Copyright © 1999 by Brian Fagan.

Published by Basic Books,
A Member of the Perseus Books Group

All rights reserved. Printed in the United States of America. No part of this book may be used or reproduced in any manner whatsoever without written permission, except in the case of brief quotations embodied in critical articles and reviews. For information, address Basic Books, 10 East 53rd Street, New York, NY 10022.

FIRST EDITION

Designed by Heather Hutchison

A CIP catolog record for this book is available from the Library of Congress
ISBN 0-465-01120-9

99 00 01 02 ❖/RRD 10 9 8 7 6 5 4 3 2 1

For Lesley
When I see you coming my heart beats
harder and I reach for your hand . . .
Or if you prefer: light blue touch paper—and stand clear

Contents

PART THREE
CLIMATE CHANGE AND
THE STREAM OF TIME

Author's Note

All measurements in this book are given in metric units. Dates are given in years before the present or in years A.D./B.C., following the most commonly used convention. All radiocarbon dates after 6000 B.C. have been calibrated with tree-ring chronologies.

Nonsailors should note that wind directions are described, following common maritime convention, by the direction they are coming *from*. A westerly or west wind blows *from* the west, and northeast trade winds *from* the northeast. It is surprising how many people are unaware of this common usage! Ocean currents, however, are described by the direction they are flowing *toward*. Thus, a westerly wind and a westerly current flow in opposite directions.

Place names are spelled according to the most common usages. Archaeological sites are spelled as they appear most commonly in the academic literature.

Preface

The whole earth is the sepulcher of famous men; and their story is not graven only on stone over their native earth, but lives on far away, without visible symbol, woven into the stuff of other men's loves. For you now it remains to rival what they have done.

—Thucydides, "The Funeral Oration of Pericles"

The damage wrought in California in 1997–1998 was not as severe as the winter storms of 1995 and 1997, but it was bad enough. We learned about El Niño long before it arrived. Satellites and computer models showed us a rapidly swelling blob of warm water—always red, like a pustule—in the western Pacific moving eastward along the equator. We were mesmerized by this expanding lesion on the earth and bombarded with predictions of approaching doom. *Time, Newsweek,* and newspapers around the world ran features on coming droughts, floods, and severe storms. The World Wide Web buzzed with dire warnings; California politicians orchestrated carefully stage-managed conferences to discuss preparations for disaster relief. It was the year El Niño became a household word and a social phenomenon. "Blame it on El Niño" became a nightclub joke in California.

The apocalypse came late to the West Coast. Day after day we basked in crisp autumn sunshine, the Pacific like glass, winds calm, the temperature neither too hot nor too cold. We began to laugh at the scientists' forecasts of record rainfall and record ocean temperatures. The fall rains came on time and were beautifully spaced.

Thanksgiving and Christmas passed with clear skies and gentle breezes. Stories of hundred-year rains in Peru, of floods that swept away entire streets of the city of Trujillo, were received with smug detachment. El Niño, we told one another, had given us a miss.

Then the January and February rains arrived. Roiling southeasterly storms descended on the California coast. Hurricane-force gusts battered San Francisco Bay and closed the Golden Gate to commercial shipping. Just south of the city, the cliffs at Pacifica melted, sweeping houses to their doom. The Russian River rose so far above flood stage that parts of the small town of Guerneville were evacuated. Mudslides cascaded down on hillside cabins and homes at the hamlet of Rio Nido. Families lost everything in minutes. Massive blocks of earth still hang over some of the homes, and they may never be reoccupied.

Storm after storm blew onto the waterlogged coast. A ferocious downpour in the mountains behind Ventura in southern California sent a flash flood rushing down the river west of the city. Within minutes the floodwaters rose over the main link between Los Angeles and San Francisco, drowning cars and blocking Freeway 101 for eighteen hours. Hundreds of motorists spent the night in their cars waiting for the waters to recede. The same flood swept away the Southern Pacific railroad trestle immediately downstream. Ten days passed before coastal rail service resumed.

California had always paid a high price for unpredictable El Niños. Fourteen years earlier, when another strong El Niño also brought strong winds and intense rainfall (as did a weaker event in 1993), floodwaters and landslides caused nearly $1 billion worth of damage between Orange County and San Diego. The saturated earth buckled sidewalks and broke concrete swimming pools like eggshells. Tornadoes and waterspouts twisted across the coast, and eight people died.

The 1998 El Niño, the greatest in living memory, brought record rains to the California coast: nearly 1,270 millimeters in Santa Barbara, almost three times its yearly average. But because federal, state, and local governments had spent millions clearing flood control channels, stockpiling sandbags, and taking other precautions, the damage was less than anticipated. It was the first time the authorities had the benefit of accurate long-range weather forecasts that predicted the onslaught. Computer models and satellite images tracked the great El Niño from birth to death. Although thousands more of us lived in low-lying coastal zones, we escaped catastrophic damage because we were at least partially ready. Everyone with access to a TV set was well aware of the impending storms.

Elsewhere in the tropical world, the 1997–1998 event caused well over $10 billion in damage. Severe droughts hit Australia and Southeast Asia. Millions of hectares of rain forest went up in smoke in Indonesia and Mexico. Over 1.8 million people in northeastern Brazil received famine relief. As in every climatic disaster, the poor suffered most, especially those living in marginal environments and in countries without the resources to prepare for drought or flood or to pay for relief and reconstruction.

Spanish colonists in Peru of four centuries ago were the first to write about El Niños. They called them *años de abundancia,* years of plenty and heavy rainfall. Ocean water temperatures rose sharply. Exotic tropical fish appeared off the coast. Vegetation bloomed in the normally arid desert. All this bounty came about because a warm countercurrent, called El Niño, "the Christmas Child," occasionally flowed southward along the Pacific coast, bringing torrential rains and exotic sea life. El Niño still brings heavy rains to Peru, and tropical fish to the nearby sea, but people now dread its ravages. Population growth has turned small farming villages into mushrooming cities where slums and shantytowns crowd onto river floodplains. A strong El Niño now sweeps away bridges, houses, and roads and kills hundreds of people, leaving hunger in its wake. The coastal economy takes heavy losses as anchovy catches plummet and fresh guano production slows dramatically. All of us, and especially the poor, are vulnerable to the Christmas Child and other short-term climatic changes as never before.

An El Niño happens when a huge "plate" of warm water accumulates in the central Pacific and moves east, slackening or reversing the northeast trade winds and bringing warm, humid air to the west coast of South America. Normal weather patterns are reversed: The deserts west of the Andes can receive their entire average annual rainfall in a day, while the rain forests of Southeast Asia and Borneo turn as dry as tinder. For years scientists thought El Niño was just a local phenomenon limited to the Peruvian coast. But in the 1960s, the UCLA scientist Jacob Bjerknes linked El Niño with atmospheric and wind circulations throughout the tropical Pacific. Bjerknes showed that El Niños were global events that triggered severe droughts, floods, and other climatic anomalies throughout the tropics.

Until recently, scientists studying ancient civilizations and those specializing in El Niño rarely spoke to one another. Now they work closely together, for they realize that this once-obscure Peruvian countercurrent is a small part of an enormous global climatic system that has affected humans in every corner of the world.

We have always known that climatic anomalies—droughts, floods, temperature extremes—could put civilizations under stress. We knew the Egyptians suffered from periodic droughts, the Moche of Peru from catastrophic rains, and the Anasazi of the American Southwest from highly localized rainfall. Such vicissitudes were seen as purely local and random phenomena, which counted for little when explaining how civilizations rose or fell. If a drought or floods happened to coincide with the collapse of a dynasty or an entire civilization, this was thought to be more a matter of bad luck than anything else. Scholarly attention focused on general ecological factors and on complex social forces such as divine kingship, increasingly centralized government, and growing social inequality.

Since Bjerknes showed that El Niño was a consistently recurring phenomenon whose effects extended around the entire world, scientific perspectives have changed. We began to see that the climatic engine that produces El Niño interacts with other major climate-producing systems as part of a huge global weather machine. Each year increasingly sophisticated computer models reveal new secrets about the world weather system and about El Niño's links with other parts of this chaotic and ever changing climatic engine. We are ever closer to learning how different states of the global machine produce predictable weather conditions on local, regional, and global scales. The study of the workings of El Niño is a microcosm of how scientists are painstakingly learning how to predict global weather.

Part One of this book describes how El Niño was first identified and the progress scientists have made in defining its role in the global weather machine. For the first time, we can infer, albeit crudely, the existence of climatic anomalies in one part of the world if we know of simultaneous (but not necessarily similar) anomalies half a world away. Thus, when a strong El Niño in the tropical Pacific produces heavy rainfall in coastal Peru, we can, with reasonable accuracy, predict a simultaneous drought in northeastern Brazil and very dry conditions in Southeast Asia.

This is a hugely important development for our understanding of world history. It means that for the first time, we have the scientific data and tools to discern, in something more than crude outline, the climatic history of human civilization. We now know that short-term climatic anomalies were not mere coincidence or aberrations. There is a strong correlation between unusual climatic shifts and unusual historical events. For example, the fall of the Old Kingdom in Ancient Egypt coincided with severe droughts that ravaged the Nile Valley in 2180 B.C.; those droughts, in turn, were triggered ultimately by interactions between the atmosphere and the ocean on the other side of the world.

Part Two of this book revolves around an increasingly important central question: How do climatic events affect the course of civilization? How do droughts, famines, and floods affect a people's faith in the institutions of their society and the legitimacy of their rulers? The newly revealed evidence of history suggests that such fluctuations present a severe—and sometimes the ultimate—test. What determines whether a society passes that test, or fails?

There are only a limited number of ways societies can respond to accumulated climatic stress: movement or social collaboration; muddling their way from crisis to crisis; decisive, centralized leadership on the part of a few individuals; or developing innovations that increase the carrying capacity of the land. The alternative to all these options is collapse. The chapters in Part Two explore different variations and combinations of these four responses. For millennia, countless Stone Age peoples of remote prehistory relied on mobility and well-developed social networks for survival (Chapter 5), as the San foragers of southern Africa's Kalahari Desert do to this day. In other instances, decisive leadership paid off. The Egyptians of 2100 B.C. survived savage droughts and the collapse of central government because local leaders with close ties to the land fed their people, then remodeled divine kingship's ancient doctrines of royal infallibility to make the kings shepherds of the people in charge of an organized oasis (Chapter 6).

Other civilizations were less adaptable because their thinking was too rigid for their environments. Fifteen hundred years ago, Moche warrior-priests in the coastal river valleys of arid northern Peru poured hydrological and irrigation expertise into their field systems (Chapter 7). They ruled with rigid, centralized control and inflexible religious ideologies. Their glittering civilization collapsed in the face of drought and then in the inevitable El Niño floods. The Maya of lowland Central America developed a brilliant civilization over two thousand years ago, a patchwork of forest states that vied with one another for power and prestige (Chapter 8). Maya lords ruled over lowland rain forest with fragile soils and unpredictable rainfall. As population densities rose, the rigid-minded rulers escalated their demands on the commoners farming a devastated environment. Then a drought cycle came and knocked out a civilization already stressed to the limit.

The Egyptians, Moche, and Maya show us that the viable options are really just two: Move away or innovate—improve the yield from the land, or pack up and settle elsewhere. In Chapter 9, I describe the

Anasazi, "the ancient ones" of the Southwest, who had no illusions about their arid environment. They developed a remarkable expertise at farming in dry environments and did not hesitate to disperse into more scattered settlements when drought cycles caused crop yields to plummet. Their descendants flourish in the Southwest today.

In Part Three, I show how the same relationships between carrying capacity, population, and the legitimacy of rulers and governments still operate today. Just like animals, all humans, whether a foraging group in the Arctic, a farming community in central Africa, or an industrial city in Brazil, live by the rules of a fundamental equation that balances population density with the carrying capacity of the land. Unlike animals, we human beings can get around the limitations of carrying capacity by increasing food supplies through technology, be it the ivory-tipped seal-hunting harpoon of a Stone Age hunter, the farmer's plow, or pest-resistant corn developed through genetic engineering. But however much we may bend this fundamental equation, we cannot escape it. Short-term climatic anomalies, whether of a few centuries or a single year, test whether we are adhering to its realities. Many times we have not, as when millions died in nineteenth-century India's monsoon failures, or when the four centuries of the Little Ice Age (Chapter 10) caused periodic subsistence crises in Europe that killed thousands. Sometimes human innovation has triumphed, as it did with the introduction of the humble potato to European agriculture in the seventeenth century. It has become fashionable in some circles to believe that human innovation will always triumph, and that population, with its inevitable needs for food, space, and waste disposal, may therefore expand indefinitely. If this were true, it would mean that humanity has entered a new and unprecedented era. However, archaeologists of the future may find this belief in infinitely bountiful technology as quaint and touching as a magical faith in divine kings.

Meanwhile, the equation of carrying capacity and population has assumed global proportions. The African Sahel (Chapter 11) offers a graphic portrait of what happens when a severe drought strikes semi-arid lands crowded with too many people and cattle. The people of the Sahel cannot move, nor do they have the capital and technology to support four times their ancient population on the same arid grasslands. They are profoundly vulnerable to starvation and to minor climatic anomalies. Even if food exists nearby, political circumstances often prevent its distribution.

The great El Niños of 1982–1983 and 1997–1998 gave us a measure of the devastation that such events can wreak across the world (Chapter 12). The material destruction, by itself, does not pose great danger to humanity. But the archaeological record shows that in societies already strained by unwise management of the environment, an El Niño adds stress upon stress, sometimes to the breaking point. In such circumstances, hunger, destruction, and dislocation can undermine the people's faith in the legitimacy of their leaders and in the foundations of their society. Overpopulation and its consequences, global warming, or rapid climate change alone will not destroy our civilization. But the combination of the three makes us vulnerable to the forces of climate as never before.

Acknowledgments

This book originated in a spirited conversation about El Niño with a group of coffee-drinking friends, who had no idea what they were launching me into. I am grateful to Noah Ben Shea, Steve Cook, and Shelly Lowenkopf for helping me have the idea and for their constant, always helpful, and often irreverent comments on my literary endeavors. My greatest debt is to William Frucht of Basic Books, who has encouraged me from the beginning and exercised superlative editorial skills on the manuscript against near-impossible deadlines. Lisa Cipolla helped with the research for many chapters and was a wonderful sounding board. Many other scholars gave advice, critiqued individual chapters, supplied references, tore apart my ideas, and took me seriously when I asked stupid questions. My thanks to Mark Aldenderfer, John Arachi, Jeffrey Dean, Carol Ellick, Michael Glantz, David Hodell, William Kaiser, Doug Kennett, Chap Kusimba, Richard Leventhal, Rod Macintosh, George Michaels, Peter Rowley-Conwy, Dan Sandweiss, Chris Scarre, Stuart Smith, Chip Stanish, Gwinn Vivian, and John Zweiker. Cindy Barrilleaux edited the drafts of the first five chapters with sensitive tact and insight. Jack Scott drew the maps and plans with his usual sensitive skill. Susan Rabiner, agent extraordinaire, helped me formulate my idea and encouraged me at every stage. Lastly, thanks to Lesley and Anastasia, who endured not only El Niño but the writing of a book about it. Small wonder Lesley sometimes calls my books doorstops!

PART ONE

The Christmas Child

Olympus, where,
they say, the gods' eternal mansion stands unmoved,
never rocked by galewinds, never drenched by rains,
nor do the drifting snows assail it, no, the clear air
stretches away without a cloud, and a great radiance
plays across that world where the blithe gods
live all their days in bliss.

—Homer, Odyssey *VI*

CHAPTER ONE

The Great Visitation

This planet's different climatic zones are all related by the winds.
These invisible threads of the climate tapestry weave the deserts
and jungles, the steppes and tundra, into a cohesive whole.
—George S. Philander, **Is the Temperature Rising?**

Come late February, the Indian sun becomes hotter with the advent of spring. First, garden flowers wither. Then wild flowering trees burst forth in scarlet and orange, the silk cotton and the coral and flame trees. Brilliantly flowering flamboyants line the sides of dirt roads, defying the ever hotter sun. By late March, the hardiest species have shed their flowers and leaves, even the golden yellow-flowered laburnum that adorns so many Punjab gardens. The sun heats and scorches as the days grow longer, drying the dew before it settles. Tinder-dry brush and woodland burst into flame, filling the dusty air with thick wood ash. The dry earth cracks and fissures as shimmering heat creates quicksilver mirages on the parched fields.

Everyone waits and waits for rain. Menacing banks of dark clouds form every May afternoon on the southern horizon as the temperature falls slightly. Solid masses of locusts cover the sun. Fine dust falls from heaven. The clouds dissipate in violent winds that fell trees and blow off roofs. The winds die down as rapidly as they came, and the heat builds relentlessly. The poet Rudyard Kipling wrote in "Two Months":

> *Fall off, the Thunder bellows her despair*
> *To echoing earth, thrice parched. The lightnings fly*
> *In vain. No help the heaped-up clouds afford,*
> *But wearier weight of burdened, burning air.*
> *What truce with Dawn? Look, from the aching sky,*
> *Day stalks, a tyrant with a flaming sword!*[1]

The aching sky—Kipling's description is an apt one. The still heat is furnacelike, bringing prickly heat and a lifeless, hazy sky of leached-out blue. Then suddenly, in May or June, the black-and-white bulbuls appear, pied crested cuckoos with long tails *(Clamator jacobinus)*, newly arrived from Africa on the vanguard of the monsoon. Black clouds build again on the horizon, flashes of lightning in their midst. Thunderclaps sound. Large raindrops spatter the waiting earth, drying as they hit the ground. Then a giant thundering erupts as torrential rain cascades onto people's upturned faces. They run around wildly in the open, waving their hands, welcoming the cool and the rain.

Monsoon rains are no ordinary storms, over in a few hours. Instead, it rains and rains. Dark clouds pass over the plains and mountains, bringing shower after shower through August and September as the monsoon spends its last force against the distant Himalayas before retreating southward in autumn. The earth turns from a desert into a sea of muddy puddles. Wells and lakes fill up. Rivers overflow their banks. Mud-hut walls melt as houses collapse. The land comes alive. Almost overnight the landscape turns green as grass sprouts, crops grow, and trees acquire new foliage. Frogs croak day and night, animals breed, farmers plant their crops, and life begins anew.

The summer monsoon is the epitome of Indian life, an experience both intensely personal and deeply spiritual, the source of human existence itself. Each year the monsoon not only brings the fullness of harvest but creates life from desolation, hope from despair. The monsoon is the smell of freshly watered earth, the sound of thunder, the season when peacocks strut their magnificent plumage, the time for merriment and lovemaking. The dark clouds of the southwest monsoon are symbols of hope, their coming commemorated by Indian writers for centuries. In the late sixth century A.D., the writer Subandhu wrote: "Peacocks danced at eventide. The rain quelled the expanse of dust as a great ascetic quells the tide of passion."[2] Eight centuries later, the poet Vidyapati wrote:

> Roaring the clouds break
> And rain falls
> The earth becomes a sea.[3]

For thousands of years Indian farmers sweated through spring and early summer, watching for that climatic moment when the monsoon rains broke. An enormous folk literature surrounds the unpredictable monsoon, doggerel and proverbs about the formation of nimbus clouds, the arrival of migrant birds, and subtle changes in vegetation. The pied crested cuckoo is said to arrive on the west coast a day or two before the rains, fly inland at a leisurely pace, and then appear in Delhi about two weeks after the monsoon breaks over the Western Ghat Mountains inshore from the coast.

Proverbs from one end of the monsoon belt to the other offer folk signs of impending rain and hope of bounty. According to Ghagh, a seventeenth-century Brahmin poet: "When clouds appear like partridge feathers and are spread across the sky, they will not go without shedding rain." He also tells us, "If clouds appear on Friday and stay till Saturday, be sure it will bring rain."[4] The proverbs disguise the agony of the long wait through searing heat, the hopeful search for the building nimbus clouds, the ecstasy of the first rains. The anguish and anticipation were based in harsh reality. Until the twentieth century, much of India was but a monsoon away from disaster.

I dimly remember my father and grandfather talking about the summer monsoon. Tall, slender, the epitome of a discreet imperial bureaucrat, my grandfather had served as financial secretary of the Punjab in the early years of the twentieth century. He was high in the intricate hierarchy of India's British Raj, a member of the governor-general's Legislative Council with awesome responsibilities if the monsoon failed. For much of the year he administered the routine of tax collection and assigned precious funds for major capital expenditures like railroads and irrigation schemes. But the summer monsoon dominated his life, and the specter of famine always overshadowed his work. Every year he awaited the monsoon with eager anticipation mingled with profound apprehension. With the monsoon came a different country, which my father once described to me by quoting E. M. Forster's *The Hill of Devi:* "Now there is a new India—damp and grey, and but for the unusual animals I might think myself in England."[5]

Even in his old age, my grandfather remembered the tense weeks of waiting, the crushing heat, the shrill cries of cuckoos, the massing clouds on the far horizon. Like the village farmers in the countryside, he watched for the coming of the monsoon and waited—for abundance or hunger. He administered the lives of thousands of villagers, but he could not control or predict the natural engine of their existence.

The word *monsoon* comes from the Arabic word *mausem* (season). The monsoon is a season of rains borne on the dark nimbus clouds of summer that blow in from the southwest. A huge circulation of air determines the intensity of the monsoon. As the earth's tilt varies with summer and winter, so the monsoon circulation moves—farther north in summer, southward in winter. In summer the northern edge of the monsoon borders on the Himalayas. Winds blow across the Arabian Sea and the Bay of Bengal, bringing moisture-laden air to

Sri Lanka in May, and to the southernmost parts of peninsular India by the first week of June. The rains move steadily northward to Bombay. By mid-June they normally cover all of Gujerat, with heavy rain along the west coast and the shores of the Bay of Bengal. In a good monsoon year, rain showers continue throughout western India and Pakistan through September, and, less certainly, from the southward-retreating monsoon into November. The agricultural lives of millions of village farmers depend on this pattern of circulation from south to north and back again. If the pattern fails, less moisture, sometimes almost none, reaches the Punjab or Rajasthan. Farther south, the usually strong southwestern monsoon winds blow with less force and drop scanty rainfall inland. Even in good years, irregular rainfall patterns can play havoc with crops of all kinds.

What happened when the unpredictable dark clouds never massed on the horizon and the monsoon failed? With almost mind-numbing regularity, Indian farmers died by the tens of thousands, sometimes millions. The story of the scientific understanding of El Niño and other global weather phenomena begins with famine and the Indian monsoon.

"Famine is India's specialty. Everywhere famines are inconsequential incidents; in India they are devastating cataclysms," wrote a Victorian traveler who witnessed the horrors of the 1896 famine in southern India. Famine was endemic in India for thousands of years, until railroads and improved communications made the shipment of grain and other food supplies to hungry villages a practical relief strategy. In 1344–1345 such a severe famine affected India that even royalty starved. The famine of 1631, following the failure of the monsoon rains in 1629 and again in 1630, devastated all of monsoon Asia. Entire rural districts were depopulated as people moved elsewhere to escape hunger and died by the roadside. Millions of cattle perished. Cholera epidemics carried away entire villages. Many areas did not recover for half a century. Another major drought came in 1685–1688. A century later the famine of 1770 caused a third of

Bengal to lie "waste and silent" for two decades. An Indian army official, Colonel Baird Smith, described how food prices rose inexorably as the rains faltered in two previous years. By January 1770, fifty people a day were dying of starvation in northern Bengal. Smith saw the dead "left uninterred; dogs, jackals, and vultures were the sole scavengers." The historian Thomas Babington Macaulay described how "tender and delicate women, whose veils had never been lifted before the public gaze . . . threw themselves on the earth before the passer-by, and with loud wailings implored a handful of rice for their children."[6] The streets of Calcutta were blocked by the dead and dying.

The South Asian monsoon failed again in 1789. A year later droughts also descended on Australia, Mexico, the island of Saint Helena in the south Atlantic, and southern Africa. The Nile River fell to record lows. The Indian drought endured until 1792, interspersed with destructive rainstorms. In three days in late October 1791, 650 millimeters of rain fell on Madras. A year later at least 600,000 people in the northern Madras region starved to death as the drought returned. No one connected the famines in India and southern Africa and the summer crop failures in distant Europe to a global weather event. They lacked the observation tools to do so.

The British administration in India had to take famine and famine relief very seriously. Official commissions studied the phenomenon and collected statistics diligently. They found that major famines had occurred about every twelve years, killing fifteen million people over a forty-year period of Queen Victoria's reign. During the 1896–1897 famine, when government relief efforts were somewhat better organized (thanks in part to improved railroads), no less than four and a half million people were on some form of temporary government assistance. Hundreds of thousands more perished of hunger and epidemic disease. As every commission and every concerned humanitarian and missionary body knew, the ever fickle monsoon rains were behind these periodic famines. The monsoon was India's salvation and scourge, and it was completely beyond human control.

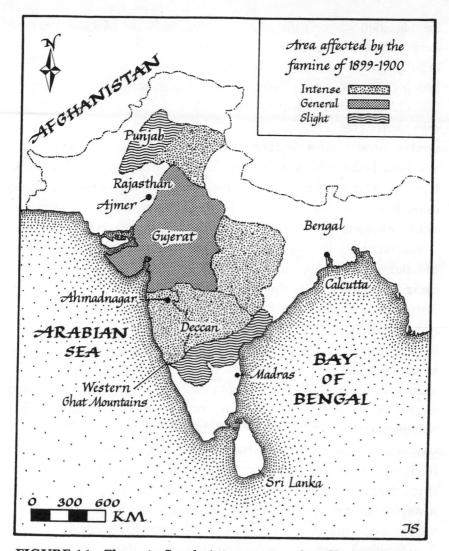

FIGURE 1.1   Places in South Asia mentioned in Chapter 1, and
the extent of the great monsoon-caused famine of 1899–1900.

The famine of 1899–1900, breaking out only two years after the
previous catastrophic drought, was the worst on record. Rainfall was
at least 27 percent below the norm of over 1,000 millimeters a year.
The dead littered villages and roadsides. The bones of famine victims

lay bleached-white in the sun. Vaughan Nash, the *Manchester Guard-ian*'s India correspondent, wrote on May 4, 1900, of "skeleton moth-ers . . . trying to keep the life in their babies—anatomies rather than living creatures; rows of emaciated children sat in silence, some of them clasping their heads in their hands and with eyes tight shut, oth-ers asleep in the dust."[7]

Over one million square kilometers of central, western, and southern India were affected. A Reuters news agency telegram to London described the fertile farmlands of the Punjab as a "vast, bare, brown, lonely desert." Of the 62 million people who were se-verely affected by total crop failure, 41.7 million lived not in native states but under British rule. A critical fodder famine killed millions of head of cattle, especially in Gujerat, where more than 70 percent perished. By March of the following year, the viceroy of India re-ported that the farmlands of the Deccan plains in the south were fast becoming a wilderness of "dismal, sun-cracked, desert-charred earth . . . sent flying in clouds of pungent dust. No water in the wells; no water in the rivers." Vaughan Nash met scores of families migrating toward government-sponsored work camps, where they would carry out manual labor in exchange for basic rations. The refugees walked in "the burning dust, with lips and throats too parched for speech, their garments often in shreds and their eyes hollow with hunger."[8] At one village in Gujerat, the fierce heat dried up the river so fast that hundreds of fish flapped in the shal-lows. Starving people from kilometers around gathered up the fish by hand and ate them. Someone in the crowd introduced cholera. Two hundred people died the first day. The villagers panicked and fled, abandoning the dead and dying. "The air became laden with the stench of putrifying bodies. . . . People suddenly sat down in the midst of conversation and rapidly sank. . . . Whichever way we turn we discover these ghastly corpses, twisted and bloated, in almost every position that agony can produce."[9]

The horrors of the 1899 famine echo down the years. The Presby-terian missionary James Inglis toured Ajmer and saw dogs fighting over the body of a child by the roadside. "I counted in one evening's journey forty dead bodies on the road, and the next day thirty-two, and the following day twenty-five."[10] In 1897 one government physician called half of India "a great charnel-house, in which countless thousands have already perished of cholera, plague, dysentery, and starvation. . . . Twenty thousand cases of cholera weekly, with a seventy-five per cent mortality, representing 15,000 deaths every seven days."[11] The situation was even worse in 1899.

The "great famine" of 1899 was documented as no Indian famine before it, thanks to photography and a sustained controversy over government and missionary relief efforts. Lord Curzon, the viceroy and governor-general, led the public appeals for humanitarian aid, but his own administration tried to spend as little money as possible on relief operations. Curzon said: "If any man is in any doubt as to whether he should subscribe, I would gladly give him a railway-ticket to a famine district. . . . He might go with a hard heart, but he would come back with a broken one."[12] The initiatives from his government, however, were grossly inadequate, especially since the authorities refused to intervene in the open market and control grain prices, which soared as crops failed. Eventually, the Indian Famine Relief Commission received millions of pounds and gifts of grain from private sources as far away as Kansas, but much of the effort was too late.

Government relief policy was, in general, devoid of any humanitarian consideration at a time when the people were weakened both physically and economically by the 1896 famine. Relief efforts began in October, long after the famine began; crop failure had become apparent in June. Curzon was stringent in his economies because of the enormous debt India owed its colonial master. About one-quarter of the Indian government's total expenditure went to pay for Britain's India Office, British officials' pensions, and interest on a rapidly increasing national debt. The excessive overhead charges levied on the Indian government by home authorities and, in turn, on village communities consumed most of India's grain surpluses in the years immediately preceding the monsoon failures of 1896–1897 and 1899–1900. Many grain shipments arrived too late and were little more than a salve for British consciences. No one knows exactly how many perished in the great famine, but it could have been as many as four and a half million people. Between 1895 and 1905, India's total population declined for ten years as a result of economic depression, repeated famines, and plague.

Missionaries called the 1899 famine "the great visitation." They preached that humanity's lot was misery and suffering. The lesson, they said, was that "natural law in its normal movement" was irrevocable and implacable; humans were helpless in the face of such emergencies as a monsoon failure. Dogma aside, many government officials, like my grandfather, worried about the constant specter of famine. How could India plan ahead to meet and mitigate such awful visitations?

The British Raj left rural India well alone, except when rapidly expanding commercial agriculture ventures aimed at overseas markets needed cheap labor. The colonial authorities were content to collect taxes and exercise administrative control while investing little in village development. Over many decades, late-nineteenth-century administrators favored a cautious policy of preventing famine rather than mitigating it. To this end, they diverted considerable resources to the building of railroads (which also helped boost India's food exports) and to the improvement of irrigation works, on the grounds that onetime capital expenditures would pay long-term dividends and could also produce revenue from tickets, freight charges, and taxes, as well as stimulate exports. In 1869, eight thousand kilometers of railroad linked Indian cities. By the end of the century, there were forty thousand kilometers of track, some of it heavily subsidized for strategic—and sometimes famine relief—purposes. For example, the government guaranteed a return on investment to a private company that built the Southern Maratha Railway in the 1880s specifically to carry grain into famine-prone areas. The strategy paid off when the collection and transport of food was better organized after 1900, for the government was able to move grain surpluses from unaffected areas into famine zones with considerable efficiency, a task that authorities called "working" a famine.

Thanks to carefully orchestrated relief policies and a slowly expanding economy, famines became a bureaucratic euphemism: "food crises." Since the early twentieth century, there has been only one famine with major loss of life–that of 1943–1944, which resulted directly from the wartime disruption of the transport infrastructure and of the economic opportunities and government relief that usually turned famine into food crisis.

While government officials grappled with famine relief strategies, British Raj scientists turned their attention to the cause of all the suffering–the monsoon that provides nearly all of India's annual rainfall. They drew on centuries of indigenous knowledge and scientific inquiry.

Merchant seaman have sailed the Indian Ocean for five thousand years. As early as 2300 B.C., King Sargon of Agade in Mesopotamia, "the land between the rivers" that is now Iraq, boasted that ships from as far afield as Dilmun (Bahrein) and Meluhha (the Indus Valley, seat of the ancient Harappan civilization) tied up at Agade's quays. Sailors in this long-distance trade down the Persian Gulf and across to South Asia must have observed the seasonal changes in ocean winds and scheduled their voyages accordingly. Mesopotamian clay tablets give tantalizing clues that seasonal departures for India in the month se-KIN-kud (February to March), when favorable winds allow easy passage to the southeastward, were underway by 2000 B.C. One thousand years later, lateen-rigged South Arabian ships traveled regularly between the Red Sea and India, coasting for days along the Arabian shoreline against the northeast monsoon. Once well to windward, the skipper would head offshore and ride the northeast monsoon to Indian shores, returning with the southwestern winds of summer. Carefully guarded knowledge of the monsoon winds passed for hundreds of years from father to son.

The secret of the monsoon cycle remained unknown to the Mediterranean countries until an Indian ship was wrecked and the skipper brought to Alexandria in Egypt. With his help, a Greek adventurer named Eudoxus of Cyzicus made two journeys from the Red Sea to India and back around 115 B.C. It was either on these expeditions or soon afterward that a Greek skipper named Hippalus worked out a strategy for much faster, direct voyaging, using the boisterous August monsoon wind to sail directly from Socotra Island at the mouth of the Red Sea to India and back within the same twelvemonth period instead of a much longer coasting journey. The Western discovery linked India with Rome, and the East African coast with Hindus and Buddhists, Sri Lanka, even distant China. The cycles of the monsoon winds became the Silk Road of the southern latitudes, a catalyst for the development of the world economic system that ultimately brought the Portuguese, British, and French to India.

"The *basadra* [summer monsoon wind] gives life to the people of the land, for the rain makes it fertile, because, if it didn't rain, they would die of hunger," wrote the Arab geographer Abu Zayd in A.D. 916.[13] Arab scholars were well aware of the rhythms of the monsoon winds. However, Arabic physics was at a loss to explain the seasonal variations in wind and rainfall, dependent as it was on a mixture of Aristotelian natural philosophy, Islamic religious belief, astrology, and folklore. Even the great tenth-century geographer al-Mas'udi was moved to remark that "the angel to whose care the seas are confided immerses the heel of his foot into the sea at the extremity of China, and, as the sea is swelled, the flow takes place."[14]

Seven centuries later, new requirements for long-distance navigation by Western nations put the study of monsoon circulation on a firmer scientific basis and prompted the first tentative studies of global climatic patterns. In 1666 the Royal Society of London prepared *Directions for Sea-Men, Bound for Far Voyages,* which contained precise instructions for the collection of data on winds and currents. The great astronomer Edmund Halley (1656–1742) used observations by dozens of seamen to prepare the first meteorological flowchart of the tropical oceans of the world. His map depicted the trade wind zones and monsoon circulations, but only in the most general terms.

In 1686 he first advanced the idea that global winds follow a consistent pattern, as part of a general circulation of air over the earth. Halley argued that differential heating of land and sea produces trade winds and monsoon circulations. He theorized that the Indian monsoons result from such heating effects and are a regional modification of the trade wind circulation. He wrote: "In *April* when the sun begins to warm those Countries to the North, the S.W. *Monsoon* begins, and blows through the Heats till *October*."[15]

Halley argued for a physical relationship between atmospheric pressure, temperature, and wind. But neither Halley nor other scientists of his day examined the ways in which atmospheric pressure and pressure variations affect wind circulation. In 1746 the Berlin Academy of Sciences went so far as to offer a prize for the best research on the laws governing air in motion, but the resulting equations were not applied to general circulations of the air around the rotating earth until the mid-nineteenth century.

During the early nineteenth century, the German explorer and scientist Alexander von Humboldt approached the monsoon problem from a different perspective. Von Humboldt was an innovative thinker, one of the first scholars to think about environmental questions on a world scale. He examined temperatures at various locations in Europe and North America and found significant differences along the same circles of latitude. Unlike earlier scientists, he argued that land-sea distributions played a vital role in modifying global wind circulations, which he regarded as the major agent of the world's climates. In 1817 von Humboldt started recording widely separated temperature distributions by using isotherms, lines that join points with similar temperatures. He adapted this recording method from the common use of isoclines to mark magnetic declination on nautical charts. Humboldt's chart showed that "the foremost effect on the climate of a place stems from the configuration of the continents surrounding it. These general causes are modified by mountains, state of the surface, etc. . . . which are merely local causes."[16]

In 1830 the German meteorologist Ludwig Kämtz constructed a global temperature chart. Using more complete observations than Humboldt, he proposed that the monsoon was controlled by differential heating and cooling of the land and sea and, just as important, by the deflecting force of the earth's rotation.

The pace of observations and research accelerated as investigators realized that latent heat in convective currents of moist air could release torrential rainfall. The American meteorologist Matthew Fontaine Maury (1806–1873) devoted his life to preparing wind and sailing charts for the world's oceans. His *Explanations and Sailing Directions to Accompany the Wind and Current Charts* of 1854 incorporated hundreds of ship observations to demonstrate the circulation pattern of the Indian monsoon and was a major advance on Halley's pioneering research of nearly two centuries earlier. As a result of Maury's work, several seafaring nations developed sailing handbooks designed to shorten the length of voyages to India and back. Maury also used 11,697 observations along the coast of the Bay of Bengal to conclude that the southwest monsoon moved southward in summer at a speed of about twenty-four to thirty-two kilometers a day.

Maury and his contemporaries realized that the study of the monsoon depended on accurate observations from many locations over long periods of time. Yet, until 1875, observations of Indian weather conditions were unsystematic and unreliable, despite the dedicated efforts of a few scientists and military officers. Rightly concerned about the safety of its merchant ships, the East India Company directed most of its research from the late eighteenth century toward studying the tropical cyclones that ravaged the Bay of Bengal every summer. The monsoon was ignored until a disastrous famine in 1866 led to the founding of the Indian Meteorological Service nine years later. Its first director, Henry Blanford, organized throughout the continent an efficient observation network that coincided to a great extent with the area of the southwest monsoon. Working with a skeleton staff, he and his successors tried to establish the causes of the southwest monsoon and the factors that triggered its torrential rainfall. From the beginning the Meteorological Department strove to develop seasonal forecasts of monsoon rainfall as a means of preparing for famines. Blanford organized a system of daily weather reports, sent by telegram to headquarters from all parts of India. By 1888 the department was furnishing daily weather forecasts.

The early forecasters were preoccupied with the dramatic onset of the summer monsoon, a memorable experience for everyone who had suffered through weeks of torrid heat. Wrote Colonel Edward Tennant of the East India Company in 1886: "The sky, instead of its brilliant blue, assumes the sullen tint of lead . . . the days become overcast and hot, banks of cloud rise over the ocean to the west. . . . At last the sudden lightnings flash among the hills, and shoot through the clouds that overhang the sea, and with a crash of thunder the monsoon bursts over the thirsty land."[17]

With their attention turned to the heavens by this awesome spectacle, the forecasters used two approaches to predict monsoon rains— the correlation between rainfall variations and sunspot cycles, and atmospheric circulation. At first, sunspot research in India appeared to show a direct relationship between monsoon rainfall and the sunspot cycle. Blanford himself analyzed sixty-four years' worth of rainfall readings from six Indian stations and found that minimum rainfall readings "somewhat anticipated" cycles of low sunspot activity, and vice versa. The British Astronomer Royal Sir Norman Lockyer, famous for his eccentric research on the Egyptian pyramids, was so impressed by this initial research that he wrote: "Surely in meteorology, as in astronomy, the thing to hunt down is a cycle."[18] Many researchers agreed that variations in solar activity affected the intensity of the monsoon as a whole. They felt, however, that the same activity had no effect on the geographic distribution of rainfall, a key factor in the extent of monsoon-caused droughts.

Like other sciences, meteorology can easily become preoccupied with local observations, what the nineteenth-century Austrian meteorologist Julius Hann called the "church tower politics" of observation—the distance one can see from atop a church tower. However, the invention of the telegraph in 1843 allowed observers to send temperature, rainfall, and pressure readings to one another in a few hours and to track severe storms as they moved over Europe. Many countries began to set up networks of observation stations after a savage gale destroyed a French fleet in the Black Sea in 1854. The tragedy could have been avoided had the telegraph alerted fleet commanders to a storm that had already caused destruction farther west.

By the 1880s and 1890s, as more scientists realized that church tower observations had a much wider context, they became interested in global patterns of atmospheric circulation. During these two decades, European meteorologists studied the seasonal movements of the Atlantic Ocean's major pressure centers, giving them names like the Icelandic Low and the Azores High. Norwegian scientists, tracking the movements of air masses, invented the term "front" to define the lines where warm and cold air clash. On the other side of the world, Henry Blanford also extended his interests to atmospheric circulation. In 1880 he showed that there was compensation of mean barometric pressure over India and Russia in winter. Blanford then took this hypothesis still further, arguing that "unusually heavy and especially late falls of snow in the North-Western Himalaya" were followed by "deficient summer rainfall on the plains." The lower snow line in the mountains, he speculated, condensed the lower levels of the atmosphere and cooled the land, thereby weakening the monsoon.[19]

Forecasts based on this hypothesis proved to be reasonably accurate in 1883, so research continued in the hands of Blanford's successor, Sir John Eliot. Eliot studied the relationship between Indian monsoon rainfall and changing barometric pressure over the southern Indian Ocean. He argued that the "burst" of the monsoon came from the advance of humid current from the equatorial zone of the southeast trade winds south of the equator. "The monsoon rains are due to the invasion of this current," he wrote. Despite page after page of detailed justification for their forecasts, Eliot and his colleagues met with little success after 1883. Matters came to a head when they failed to forecast the terrible drought of 1899. As a result of an outcry in the press, their predictions were no longer published in the media.

John Eliot's successor was Sir Gilbert Walker, the greatest of all directors of observatories in India. Unlike his predecessors, Walker was not a meteorologist. From 1895 to 1902, he was a senior wrangler in mathematical physics at Cambridge University, where his specialty was electrodynamics. His interests and publications ranged over electromagnetism, games and sports, even the flight of birds. He had such a passion for boomerangs and other primitive hunting devices that he earned the nickname "Boomerang Walker" from his Cambridge friends. This modest and liberal man was the epitome of the English gentleman.

Walker was appointed to the Foreign Service in 1903, served for six months as Eliot's meteorological assistant, and then assumed charge of the grossly understaffed weather service late in that year, at a time when accurate forecasts of monsoon rainfall were the director's most important concern.

Walker's lack of meteorological experience turned out to be a godsend. He was an expert statistician who believed that "what is wanted in life is ability to apply principles to the actual cases that arise." For twenty years in India, and during an additional three decades after his retirement, this remarkable scientist used statistical methods and thousands of local observations to establish the relationships between the complex atmospheric and other conditions that affect monsoon rainfall. A brilliant administrator and organizer, Walker soon expanded his work far beyond India. He established that monsoon droughts do not result from human environmental modification, such as deforestation. In a memorable paper published in 1910, he examined rainfall data from India and the Nile Valley and concluded there was no evidence that India's disastrous droughts had been caused by permanent climate change. In the same year he wrote in the *Memoirs* of the Indian Meteorological Service: "The variations of monsoon rainfall ... occur on so large a scale [that we can assume they are] preceded and followed by abnormal conditions at some distance."[20] With these words, he turned his attention to the complex interrelationships between the monsoon and global atmospheric circulation. Three years later he was able to show that increased sunspot activity could intensify existing monsoon conditions but did not play a major role in monsoon failure.

Gilbert Walker was a tireless scientist with a passion for detail and statistical calculation, but he maintained a wide grasp of much larger problems. Like his predecessor, Sir John Eliot, he looked far beyond India for global predictors—complex associations of widely separated atmospheric and weather events that could cause drought in India. By 1908 he had developed a forecasting formula in the form of a regression equation that drew on years of rainfall observations, and also on global phenomena observed by other scientists. The British astronomers Norman and William Lockyer had identified a complex pressure seesaw between South America and India in 1902. Walker was also aware of research by H. H. Hildebrandsson, who had observed an opposition of barometric pressure between Buenos Aires and Sydney, Australia, and of recent statistical studies of weather anomalies in the northern hemisphere. In a series of papers (his most important), "Correlations in Seasonal Variations of Weather," published in the *Memoirs* in 1923–1924, Walker identified what he called "strategic points of world weather." He wrote: "We can best summarize ... the situation by saying there is a swaying of press[ure] on a big scale backwards and forwards between the Pacific Ocean and the Indian Ocean, and there are swayings, on a much smaller scale, between the Azores and Iceland, and between the areas of high and low press[ure] in the N. Pacific."[21]

Walker named the most important of these swayings the Southern Oscillation. "By the Southern Oscillation is implied the tendency of pressure at stations in the Pacific ... , and of rainfall in India and Java (presumably also in Australia and Abyssinia) to increase, while pressure in the region of the Indian Ocean ... decreases."[22] By 1924, when he retired from the Indian Meteorological Service to become professor of meteorology at Imperial College in London, Walker was publishing charts of the Southern Oscillation in both summer and winter based on data from an enormous network of observation stations between Africa and South America. His charts clearly demonstrated relationships between pressure, temperature, and rainfall in the Pacific and Indian Oceans, between the intensity of monsoons and earlier, changing pressure conditions thousands of kilometers away. Still, Walker could not predict monsoon failure. As he himself admitted, such forecasts would be successful only in years of strong statistical relationships. Ever judicious, he preferred the term "foreshadowing" to "forecasting," what he called a "vaguer prediction." As his successor, Sir Charles Normand, observed prophetically in 1943, Walker's worldwide surveys offered more promise for weather prediction in other regions than for monsoon rainfall, where his research began.

Sir Gilbert Walker discovered the relationship between the Southern Oscillation and Indian monsoon rainfall, but his regression formulas, while considerably better than guesses, were only slightly better than those based on probability tables.[23] Charles Normand verified eighteen years of monsoon forecasts (1931–1948) based on the formulas from Walker's research. For years of deficient rainfall alone, 66 percent of the forecasts were wrong. Normand questioned whether these particular predictions were any use at all. However, he favored their continuation "if only to keep the subject alive and in the hope that ideas for progress will emerge."[24]

Walker's attempts to establish linkages between pressure anomalies, rainfall, and temperature in widely separated parts of the world were challenged or ignored by many of his contemporaries, despite their interest in such links. Unlike Walker, many of them relied on qualitative assessments or just plain guesses. His obituary in the *Quarterly Journal of the Royal Meteorological Society* for 1959 commented that "Walker's hope was presumably not only to unearth relations useful for forecasting, but to discover sufficient and sufficiently important relations to provide a productive starting point for a theory of world weather. It hardly seems to be working out like that."[25]

Half a century later, Gilbert Walker is remembered as one of the great heroes of global climate and El Niño research. The Pacific atmospheric circulation that links the Southern Oscillation with sea surface temperatures now bears his name: the Walker Circulation.

CHAPTER TWO

Guano
Happens

*The Peruvian El Niño is not a unique occurrence, but is rather
merely a striking example of a wide-spread occurrence.*
　—*Warren S. Wooster, "El Niño," California Comparative
Oceanic Fisheries Investigations,* Reports

On a windy day in the 1850s, the inhabitants of Pisco, on Peru's
southern coast, could smell the rotting guano on the precipitous
slopes of the volcanic Chincha Islands fifteen kilometers offshore.
Few of them ever visited this terrible place, where millions of
seabirds nested and left mountains of droppings atop the high cliffs.
But they could not have avoided hearing stories of the horrors or seeing foreign laborers herded into crowded boats for a short voyage
from which few returned. Across the water, hundreds of indentured
laborers, convicts, army deserters, and coolies were worked to death,
laboring twenty hours a day without respite under the eyes of ruthless
supervisors and fierce dogs. They dug deep trenches into the hardened guano with picks and shovels, earning a few pesos for each ton
they extracted. Dilapidated ships were moored alongside the precipitous cliffs. Gaunt, exhausted men tipped wheelbarrow loads of ce-

mented bird droppings into filthy canvas chutes into the holds below. Mainlanders could see lung-clogging dust hovering over the island on still days as the toxic smell of ammonia filled the air. Blistered hands and cut legs turned septic. The sick and injured became wheel-barrow hands, coughing from chronic respiratory ailments or bent over with gastrointestinal disorders. Each morning a burial detail car-ried off the dead and interred them in shallow graves, where raven-ous dogs dug up the corpses and gorged themselves on fresh meat. No one bothered to rebury the rotting bodies.

Bird guano was one of the most valued commodities in the mid-nineteenth-century world. A wealthy and politically influential con-sortium of foreign and Peruvian businessmen made millions from the export of this nitrogen-rich natural fertilizer to Europe and North America. They callously killed thousands of miners in the process.

The story of El Niño also begins with this unlikely substance—guano. In 1803 Alexander von Humboldt and the French botanist Aime Bonpland traveled through the coastal desert of Peru. Their journey took them through one of the driest environments on earth, where lit-tle or no rain falls for years on end. Only mountain runoff from the Andes waters the coastal river valleys. To their astonishment, the ex-plorers observed bountiful crops growing along the rivers. They dis-covered that the coastal farmers used seabird droppings as fertilizer to produce luxuriant plant growth, so much so that the Peruvians had a proverb: "Guano, though no saint, works many miracles."[1]

For thousands of years, countless guanay cormorants, gulls, peli-cans, and other seabirds had weaved and whirled over the cold in-shore waters of the Pacific, where the Humboldt Current (named af-ter our explorer) swept northward along the coast of South America. The birds feasted off teeming shoals of anchovies and other fish, roosting and nesting on small, rocky islands just offshore from the mainland. Their droppings, estimated at forty-five grams per bird a day, accumulated steadily on the craggy rocks. Birds produce semi-solid urine with a high concentration of the nitrogenous compound uric acid, a stable and insoluble substance. The dry climate preserved the guano as a stinking, phosphate-rich, yellowish concretion. By von Humboldt's time, after accumulating for more than twenty-five hundred years, guano covered some of the Chincha Islands to a depth of more than thirty meters (for map, see p. 122).

As fertilizer, Peruvian seabird excrement had a long and honorable history. The Moche Indians of northern Peru mined the Chincha Islands for guano as early as A.D. 500 and built a flourishing agricultural state on its nitrogen content. Centuries later the Incas so valued bird droppings that they divided portions of the islands between different peoples and communities. *Huano,* like gold, was considered a gift of the gods. The death penalty awaited anyone caught killing nesting birds.

Despite von Humboldt and Bonpland's discovery, the remarkable properties of Peruvian guano remained unknown to the outside world until the 1830s. A few casks eventually reached Europe and North America, but it was not until 1840 that field experiments showed guano was vastly superior to manure. Moreover, unlike manure, a full guano sack was light enough to be carried up steep slopes with ease.

In 1840 and 1841 a consortium of British, French, and Peruvian businessmen acquired a monopoly on guano exports and loaded almost eight thousand tons of it on Europe-bound ships. The stench of the first shipment was reportedly so foul that the inhabitants of Southampton, England, took to the hills in disgust. Considering the poor sanitation in European cities of the day, the smell must have been truly awesome. So were the profits. The consortium could land a ton in Britain for about $30 and sell it wholesale for $90. The guano trade mushroomed into big business almost overnight. For the next forty years the Peruvian government earned most of its foreign exchange from selling bird excrement in Europe and North America. With its uniquely high nitrogen content, Peruvian guano had no rivals until manufactured fertilizers came into widespread use for

turnips and other root crops in the 1860s. Farmers on the Chesapeake Bay praised guano as thirty-five times more effective than barn
manure. Rutabaga and turnip crops in some parts of Maine and
Massachusetts quadrupled when fertilized with guano.

For a quarter-century the feces trade, with all its horrors, sustained
Peru, reaching a peak in 1855, then dropping steadily, much to the
concern of the wealthy landowners who kept a tight grip on the
guano monopoly. Their ruthless mining operations rapidly stripped
the offshore islands. In the late 1860s, word that Peruvian guano supplies were nearing exhaustion swept Europe and North America. The
rumors coincided with local newspaper stories of beaches littered
with seabird carcasses. Millions of birds died within a few months as
torrential rains battered the normally arid coastline. Sheets of rainwater washed many cubic meters of guano from the steep-sided islands,
further contributing to the sense of impending ruin. The guano authorities tried to reassure their customers that bird droppings were a
renewable resource, that the seafowl would return. Return the birds
did, as they had in previous years after many died. But guano exports
never again matched the boom years of the 1840s and 1850s. Nevertheless, more than twenty million tons of guano left Peru for Europe
and the United States between 1848 and 1875, with a value of some
$200 million in present-day currency.

At the time when guano was one of the staples of the Peruvian economy, local scientists puzzled over the seabird mystery. No one then
understood the close interrelationships between animals and their environments, nor did they comprehend the complex interactions between fish and ocean-floor nutrients off the desert coast. They knew
there were unexplained "years of abundance" with warmer than
usual ocean temperatures and heavy coastal rain. A traveler in 1891
wrote home of a transformed coast in such a year: "The sea is full of
wonders, the land even more so. First of all the desert becomes a garden. . . . The soil is soaked by the heavy downpour, and within a few
weeks the whole country is covered by abundant pasture. The nat-

ural increase of flocks is practically doubled and cotton can be grown in places where in other years vegetation seems impossible."[2]

The scientists eventually talked to the coastal fisherfolk and sea captains. They found that some skippers had noted unusual currents and warm seawater conditions off the coast in ship's logs going back to 1795. In 1891, which was just such a "year of abundance," Luis Carranza, president of the Lima Geographical Society, contributed a short paper to the society's *Bulletin* in which he drew attention to an occasional countercurrent that flowed from south to north along the coast. His contribution drew little notice. The following year, the sea captain Camilo Carrillo published another paper in the same *Bulletin*. He wrote: "The Paita sailors, who frequently navigate along the coast in small craft, either to the north or the south of that port, name this counter-current the current of 'El Niño' (the Child Jesus) because it has been observed to appear immediately after Christmas."[3] In 1895 the Peruvian geographer Victor Eguiguren theorized that heavy rains and flooding around the northern town of Piura were linked to the El Niño countercurrent.

Then Frederico Alfonso Pezet addressed the Sixth International Geographical Congress in Lima in the same year. His subject was El Niño. In front of a large and learned audience he said: "I wish, on the present occasion, to call the attention of the distinguished geographers here assembled to this phenomenon, which exercises, undoubtedly, a very great influence over the climate conditions of this part of the world."[4]

El Niño thus entered the world of international science.

For years scientists everywhere assumed that the Peruvian countercurrent was a local phenomenon that merely disrupted the even tenor of ocean life along the Pacific shore. *Años de abundancia* brought torrential rains, strange tropical fish species, and even water snakes, bananas, and coconuts to Peru, but they seemed to affect only inshore waters. The American biologist Robert Murphy reported that during the 1925 El Niño, "one afternoon, rising along the beach, I flushed a fair-sized alligator, which hoisted its tail and dashed into the sea."[5] El Niño became one of those minor meteorological curiosities that affected a desert land far away and on a far smaller scale than the destruction wrought by Indian monsoon failures.

Every year extreme weather strikes some part of the globe. El Niño seemed like one of those irregular, purely local incidents that crop up year after year, sometimes causing little more than a ripple in the coastal fisheries or a localized flood on Peruvian north-coast rivers. Some years, however, the Christmas Child descended with wrathful vengeance: Ocean temperatures rose so high that all the anchovy moved away and millions of seabirds perished.

In 1925 the weather went crazy. In January fisherfolk at Talara in extreme northern Peru reported unusually warm sea temperatures. The ocean temperature rose by 6.6 degrees Celsius in ten days and kept on rising. The warming waters moved southward. By March the Pacific temperature off the Moche River had risen to twenty-eight degrees Celsius, sixteen degrees above normal, before settling down to a steady twenty-two degrees over more than eleven hundred kilometers of the Peruvian coastline. Some cooling began in April, but warmer than usual temperatures would persist right through 1930. As sea temperatures rose, the easterly trade winds faltered, then reversed. At least twenty-four million cormorants and other seabirds died of starvation as the anchovies moved away into cooler, nutrient-rich waters. Seabird corpses littered beaches and offshore islands. So much sea life perished and decomposed on the ocean floor that foul-smelling hydrogen peroxide blackened ship's hulls and discolored houses by the water's edge.

Meanwhile, air temperatures rose over the land. Huge black clouds built over the western horizon off the Moche River, then moved inshore, bringing massive thunderstorms. Cloudbursts saturated the coastal plain, turning dry water courses and ravines into torrents. In early March, 226 millimeters of rain fell in three days. By the end of the month, the city of Trujillo, whose normal rainfall was 1.7 millimeters, had received 396 millimeters. Walls of water swept down the Moche River, flooding roads and taking out the Trujillo

railroad bridge. Thousands of acres of farmland and countless irrigation canals vanished under a sea of mud and hillside debris. No one
knows exactly how many people died in the 1925 El Niño, but they
number in the hundreds, if not thousands.

At the time, Trujillo was a quiet farming town, remote from the
center of government at Lima, five hundred kilometers to the south.
The floods cut the community off from the outside world for weeks.
Fish catches had plummeted, and the guano industry had come to a
virtual standstill. All the farmers could do was to clear their lands and
restore their irrigation canals, hoping for a good harvest in a few
months. Hundreds of people perished from starvation. Many more
undoubtedly suffered acute hunger or malnutrition.

How can we explain such sudden climatic savagery? For many years
Peruvian scientists puzzled over the inconsistencies in their coastal
climate, normally so predictable and benign. They slowly pieced together a portrait of El Niños large and small, of the local currents and
ocean upwellings that made the Peruvian coast one of the richest marine environments on earth.

Peru's coastal people have lived in a unique setting of coastal
desert, bountiful ocean, and high mountains for at least ten thousand
years. The Andean mountain range runs parallel to the long Pacific
coast from three degrees south of the equator southward for over fifteen hundred kilometers. The cordillera rises abruptly from the
coastal plain, one of the driest environments on earth. Some of Peru's
greatest civilizations, like the Moche of A.D. 500, flourished along this
narrow coastal strip, in valleys that reach an altitude of four thousand
meters a mere one hundred kilometers from the coast. The mountain
ranges and the cool Humboldt Current close offshore determine the
coastal climate. The prevailing southeast trades carry most Pacific
water northward toward the equator. Most years little or no rain
reaches the arid coastal plain. All surface water comes from mountain
runoff, which flows down coastal river valleys to the ocean (see Figure 7.1 on p. 122).

The Andes receive most of their rain from the Amazon Basin to the east during the southern summer, November to April. Amazonian water vapor arrives on the prevailing equatorial easterlies and falls as rain on the exposed windward escarpments of the mountain ranges. Little of this rain reaches the western slopes. The Peruvian coast is watered by precipitation from the other side of the continent.

An equatorial trough of relatively low pressure lies between the northeast and southeast trade wind belts, migrating northward and southward according to the season. In summer the trough moves south. The prevailing easterlies over the Amazon Basin interact with the unstable equatorial trough and bring rain to the high Andes. These rain-bearing winds are the main source of mountain runoff for the Pacific coastal valleys. During the winter months strong upper-level westerlies keep moisture at bay. Deep ice cores from Andean glaciers tell us that this seasonal pattern has remained constant since rising post–Ice Age sea levels stabilized around 5000 B.C., and perhaps since the end of the Ice Age ten thousand years earlier.

This same rainfall pattern occurs throughout the fifteen degrees of tropical latitude straddled by the Peruvian mountains until the drier high-altitude *altiplano* grasslands around Lake Titicaca in southern Peru and Bolivia. The seasonal rhythm is remarkably consistent, except in El Niño years.

The Christmas Child, unpredictable and of highly variable strength, is the trickster. Most episodes are of relatively little consequence, bringing occasional periods of torrential rainfall, much warmer sea temperatures, and unfamiliar tropical fish to coastal waters. However, once a generation or so, an exceptionally strong El Niño episode causes drastic ecological and climatic changes along the coast, with potentially catastrophic consequences for farmers and fisherfolk alike.

The anchovy fishing grounds lie close off the arid coastline, in an area where the sea bottom drops precipitously. There is no continental shelf to trap nutrients as there is off Europe and the eastern United States. The easterly trade winds, pushing the surface water to the west, cause cold, nutrient-rich water to well up from the depths. This profile—prevailing offshore winds, narrow continental shelf (or none), and consequent upwelling of very deep water to the surface—occurs in only four other places in the world: off the California coast and at three locations in Africa—Mauritania, Namibia, and Somalia. These five areas, 0.1 percent of the ocean's surface, account for about half the world's commercial fish catch.

Peru's anchovy fishery has sustained human life along the coast for over ten thousand years, especially with protein-rich fish meal, a staple for millennia. Before commercial fishing depleted the anchovy shoals, the swarms were almost unbelievable. The archaeologist and diplomat Ephraim Squier landed on the north coast in 1865. His small open boat rowed through a sluggish swell, passing through "an almost solid mass of the little fishes . . . which were apparently driven ashore by large and voracious enemies in the sea. . . . The little victims crowded each other, until their noses, projecting to the surface, made the ocean look as if covered with a cloak of Oriental mail. We could dip them up by handfuls and by thousands." Squier described a mile-long belt of fish close to the shore. Women and children were scooping them up "with their hats, with basins, baskets, and the fronts of their petticoats." He also admired the Indian reed canoes, with tapered ends, which could carry several people or large fish catches through the breakers. "They are extremely buoyant and when not in use are taken ashore and placed erect to dry."[6] The ancient Moche used the same vessels and depicted them in clay.

Two north-flowing cold ocean currents dominate the coastal waters off Peru. The Humboldt Current is between five hundred and six hundred kilometers wide and can reach a depth of seven hundred meters. This ocean river can flow so strongly that nineteenth-century coastal steamers charged 10 percent more for a southbound passage, to pay for the extra fuel they consumed. The Peru Coastal Current, no more than one hundred to two hundred kilometers wide, runs close to shore and rarely exceeds a depth of two hundred meters. The south-flowing Peru Undercurrent moves the water beneath these two streams. Southeast trade winds prevail over the coast, blowing parallel to the shore or offshore. The rotation of the earth causes the Coriolis Force, which deflects surface water to the left of the wind direction in the Southern Hemisphere. As a result, surface water moves offshore along the entire Peruvian coast. As these waters flow away from the coast, subsurface water moves upward slowly to replace the displaced mass. The rising waters bring with them a dense concentration of nutrients, stimulating truly remarkable biological productivity.

This bounty seems inexhaustible until a strong El Niño dramatically alters the normal coastal upwelling. Warm, nutrient-poor water from near the equator north of Peru now travels southward along the coast on a strong countercurrent. The sea temperature difference can be as much as ten degrees Celsius. This less dense, less saline water overrides the much cooler Coastal Current water with a nutrient-poor layer up to thirty meters deep. As the El Niño strengthens, the coastal winds slacken, giving the surface waters less of an offshore push. Upwelling weakens or even ceases altogether. Even if the winds continue to blow and upwelling persists, the nutrient-poor upper layers are so thick that the upwelled water is also low in nutrients.

Photosynthesis now drops dramatically in the coastal current system. The anchovy shoals either concentrate in pockets of cold water close inshore, disperse into deeper cold water much farther offshore and out of canoe range, or simply die. Thousands of fish also are eaten by warm-water predators like horse mackerel, which move closer inshore with the countercurrent. In either case, anchovy catches fall dramatically. Other marine organisms such as squid and turtles also starve. So do fish-eating birds such as cormorants, gannets, and pelicans, which feed off anchovies almost exclusively.

As it has done for thousands of years, a strong El Niño hits the coastal people below the economic belt from several directions. While fishing communities could subsist on the unfamiliar species that entered coastal waters, there was much less food to go around for people living inland. Fish meal production plummeted, with only a fraction of normal supplies moving inland and to the distant highlands. At the same time, torrential rain and flooding inland caused massive erosion and destroyed irrigation schemes. Strong El Niños brought catastrophe and suffering and destroyed villages, cities, and even entire civilizations.[7]

While the Peruvians dug out from the 1925 El Niño, the Indian monsoon expert Sir Gilbert Walker was settling in as professor of meteorology at the University of London. His long and distinguished career in India had armed him with roomfuls of statistics about atmospheric pressure in every part of the world. Though he was no closer to predicting monsoons, he understood they were part of far larger, and still mysterious, global climatic phenomena.

In an era long before computers and electronic mail, he enlisted hundreds of weather technicians and clerks for the laborious task of computing the statistical correlations between climatic variables collected from different observation stations around the world. These variables included surface pressure, temperature, rainfall, and sunspots. In a series of tightly argued papers published during the 1920s, Walker concluded that when atmospheric pressure was high in the Pacific Ocean it tended to be low in the Indian Ocean, from Africa to Australia, and vice versa. This "Southern Oscillation" seesawed back and forth, changing rainfall patterns and wind directions over both the tropical Pacific and Indian Oceans.

The Southern Oscillation is Gilbert Walker's most lasting scientific legacy. Inadvertently, he discovered the parent of El Niño.

Walker's fanatical preoccupation with statistical correlations of meteorological phenomena across the globe caused him to search ever harder for better predictions of monsoon failure, to the point that he sometimes ignored the limitations of his approach. He was at the mercy of his figures, without access to the sophisticated computer-based modeling that is routine today. His results were controversial from the beginning. Many colleagues rightly questioned the long-term stability of his pressure oscillations, arguing that his correlations came from observations collected over relatively short periods of time. These relationships might change, they argued, but Walker's calculations would always produce a few chance correlations, even if the data consisted of nothing more than random numbers. Others questioned his free use of long-distance correlations, such as that between the pressure in the center of the Southern Oscillation and at Charleston, South Carolina. Walker replied to his critics disarmingly and prophetically: "I think the relationships of world weather are so complex that our only chance of explaining them is to accumulate the facts empirically."[8] He also pointed out that he had only surface observations to work with and none from the atmosphere.

In the end, Gilbert Walker failed to predict monsoons. His greatest achievement was to recognize that the ocean plays an important role in the mechanisms of the Southern Oscillation. But he lacked the sea surface temperature and subsurface data to confirm this all-important relationship. In quiet, gentlemanly despair, he devoted much of his retirement to a musical passion—improvements in the design of the flute.

Professor Sir Gilbert Walker died in 1958 at the age of ninety, just as the tropical Pacific experienced an unusually powerful combination of much warmer water temperatures and weak easterly trade winds. The years 1957–1958 spawned the most powerful El Niño since 1941. By fortunate coincidence, 1957–1958 also marked the International Geophysical Year (IGY), which brought together scientists from both sides of the Iron Curtain and all parts of the world in a unique global cooperation to study the earth. The oceans figured prominently in these activities.

A small number of IGY scientists became fascinated with the strong pattern of anomalies over the tropical Pacific in both the atmosphere and the ocean in 1957. Their initial analyses revealed some unusual phenomena. A strong El Niño developed along the Peruvian coast. Yet ocean warming extended thousands of kilometers across the Pacific to the International Date Line and beyond, no less than one-quarter of the way around the world. Could this extensive warming be connected to El Niño? Furthermore, the easterly trade winds that normally prevail over the central Pacific were unusually light. The light winds coincided with exceptionally heavy rainfall in the same area. In the eastern North Pacific, stronger westerly winds resulted from pronounced anomalies in the atmosphere, while the low-pressure center that normally sat over the Aleutian Islands was much stronger than usual.

Among these scientists was the Norwegian meteorologist Jacob Bjerknes, who had moved to the University of California at Los Angeles while at the pinnacle of his career. As a young man, Bjerknes had studied the theory of air-mass circulation, especially fronts and cyclones. These studies led him to research the general circulation of the atmosphere in the North Atlantic. But the strong El Niño of 1957–1958 turned his attention to the tropical Pacific, with momentous consequences.

Bjerknes had a global perspective on atmospheric circulation and a penchant for looking at any form of climatic anomaly on a very broad canvas. In 1966 he boldly suggested that the seemingly unusual pattern of atmospheric anomalies throughout the Pacific was not unique. Exactly the same pattern had repeated itself in earlier El Niño episodes and would reappear in the future. At the same time, he theorized on the basis of what he called "tenuous reasoning" that the ocean warming observed in 1957 had been related to atmospheric circulation anomalies far away in the North Pacific. By this time, Bjerknes had data from two more "analogous happenings," as he called them, in 1963 and 1965. But instead of relying only on surface readings, he could call on satellite imagery to define an area of unusually high rainfall over the central tropical Pacific. In 1969 he published a landmark synthesis of years of circulation research in which he demonstrated conclusively the intimate relationship between Walker's Southern Oscillation and these anomalous patterns in the Pacific.

Bjerknes's hypothesis began with the assumption that the normal sea surface temperature gradient between the relatively cold eastern equatorial Pacific and the huge pool of warm water in the western Pacific as far west as Indonesia causes a huge east-west circulation cell on about the plane of the equator. Dry air sinks gently over the cold eastern Pacific. Then it flows westward along the equator as part of the southeast trade wind system. The western "push" that drives this movement comes from atmospheric pressure that is higher in the east and lower in the west. The air becomes warmer and moister as it moves over progressively warmer water. In the western Pacific, the moisture condenses out in the form of towering rain clouds. A return flow to the east in the upper atmosphere completes the circulation pattern.

Bjerknes named his newly discovered circulation the Walker Circulation, in honor of Sir Gilbert.

Bjerknes realized that this critical atmospheric circulation was the key link between the Southern Oscillation and the variations in sea surface temperatures in the equatorial Pacific. He reasoned that when warming occurs in the eastern Pacific, the sea surface temperature gradient between east and west decreases. The trade wind flow that forms the lower branch of the Walker Circulation weakens. However, a weaker east-west pressure gradient has to accompany the reduced trade winds. Such changes required pressure changes between the eastern and equatorial Pacific to act like a seesaw—precisely what happens with the Southern Oscillation.

The Bjerknes hypothesis, published in 1969, forged the final link between El Niño episodes and the Southern Oscillation. The great meteorologist described a highly complex set of interactions making up the El Niño–Southern Oscillation connection, known commonly as ENSO. With brilliant elegance, supported by convincing dynamic and thermodynamic reasoning, Bjerknes pulled together a number of critical elements into a new conceptual framework—the constant seesaw movements of the Southern Oscillation, the large-scale air and sea interactions that cause Pacific warming, and some much larger global teleconnections with climatic variations in North America and the Atlantic Ocean. The Bjerknes hypothesis has two important qualities: robust and logical reasoning, and a flexibility that has allowed it to accommodate the rapid accumulation of new data. Few scientists in such a fast-moving field have ever made such a lasting contribution.

Six decades after Gilbert Walker took his statistical skills to India, Bjerknes showed that ocean circulation is the flywheel driving a vast climatic engine. Bjerknes also proved that the Southern Oscillation is a powerful and often dominant global climate signal on short and unpredictable time scales ranging from a few months to several years. Death came to starving seabirds at the hands of a Christmas Child generated by a climatic powerhouse thousands of kilometers away.

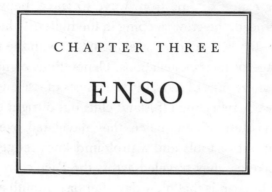

CHAPTER THREE

ENSO

The partners in this dance are the atmosphere and the ocean. But who leads? Which initiates the eastward surge of warm water that ends La Niña and starts El Niño? Though intimately coupled, the ocean and atmosphere do not form a perfectly symmetrical pair. Whereas the atmosphere is quick and agile and responds nimbly to hints from the ocean, the ocean is ponderous and cumbersome.

—*George S. Philander,* Is the Temperature Rising?

Sailing in the northeast trades of the Pacific: a steady fifteen-knot wind blowing over my right shoulder, the warmth caressing my bare shoulders, the gentle rolling as our boat cascaded across the deep blue ocean. I braced myself against the constant motion, gazed out for hours at an empty horizon and a sky of marching, puffy clouds. More than two thousand kilometers from land: day after day the same unchanging routine, no need to adjust a sail or disconnect the self-steering gear. Each day I checked the rigging for chafe as the sun set. At night the brilliant moon bathed the deck in silvery light so bright I could read a book. I marveled at the sheer predictability and continuity of it all, at a passage that took me out of the stream of time.

After a few days the passing stars were old friends. I remember thinking of the ancient Polynesian navigators who had sailed these same waters in their double-hulled canoes long before us. I imagined the low-slung canoe surging from wave to wave, bellying fiber sail straining overhead, the crew sleeping in the hulls. By the steering oar, an aged pilot stands, feet apart, swaying easily with the swell, observing the passage of the constellations. Generations of navigators like him learned the zeniths of stars, the sequences of rise and set, the subtle signs of swells reflecting off island cliffs out of sight over the horizon. With unassuming confidence, they navigated over empty seas with the simplest of tools and a profound knowledge of their surroundings. Like us, they traveled across the Pacific on a band of utterly predictable winds that blew day after day, month after month.

After two days, the peaks of Bora Bora appeared over the dawn horizon, exactly where they should have been. Like the navigators of old, I felt the profound, humbling satisfaction of a perfect landfall made on the arms of the trades.

We live in a world of seething air, which flows around us in a state of constant change. Violent hurricanes, tropical cyclones, searing desert winds, and months of steady trades: The only constant is change. The sun is another constant, rising and setting every day in an endlessly repeating cycle of birth, life, and death, alternately warming and cooling the air around us. Sunrise and sunset give such an illusion of continuity, of an unchanging world, that many ancient societies worshiped the sun. Aztec priests sacrificed human victims to the Sun God by tearing out their still-beating hearts with an obsidian knife, in the belief that such offerings would perpetuate their finite world, destined, according to legend, to end in a swarm of earthquakes. Ancient Egyptian pharaohs were the living personification of the Sun God on earth, shepherds of their people, masters of the life-giving Nile. They and their subjects lived according to *ma'at,* a spirit of harmony, justice, order, and peace. *Ma'at* was celestial harmony, a recognition of the eternal qualities of existence as demonstrated by the earth and sky.

The Egyptian world could indeed be orderly and predictable, a universe where day after day the sun rose in a cloudless sky and set into the desert hills while bathing them in a delicate rosy light. But even the narrow Egyptian sphere lay at the mercy of drought and flood throughout the Nile Valley. The pharaohs lived, all unaware, in a world dominated by the interlocking cycles of ever changing global weather.

The sun rises and sets, the air warms and cools, water evaporates, rises, cools, then falls as rain before the endless cycles begin again. Great winds chase relentlessly around the globe. Low-pressure centers form, suck high-pressure systems into their vacuum, create spinning masses of clouds and wind. The weather cycles and recycles, clashes and grows calm—all because the sun warms some parts of the spherical globe more than others. I have excavated ancient African villages close to the equator, where the sun beats directly overhead, making work impossible in the middle of the day. We would dig from dawn until midmorning, then rest in the shade until the afternoon shadows lengthened and a light wind blew over the trenches. A few months later I happened to visit Alaska's North Slope in early summer, when the sun barely sets but is always near the horizon. The temperature was only a few degrees above zero, because the sun's rays were weak. The equator is warm, the poles are cold, so solar heat flows from the tropics toward the Arctic and Antarctic. Solar energy fuels our global weather machine. Warm air leaves the tropics for the poles. Cold air from high latitudes travels toward the equator. The two-way movement forms vast convection cells that make the wind blow.

Our planet spins as it travels around the sun. The earth's rotation pushes the air in the convection cells sidewards as they are dragged by friction from the land and sea or squeezed by gravity. The Coriolis Force, named after the French mathematician Gustave Gaspard Coriolis (1792–1843), results from the earth's rotation. In the Northern Hemisphere, the path of an object appears deflected to the right, in the Southern to the left, a matter of some concern to people launching artillery shells or rockets. The sideways push given the winds by the earth's constant spinning causes the convective flows to organize in great bands, where the flow direction varies according to latitude. The jet streams high in the atmosphere and the prevailing winds on the ground both result from the Coriolis Force.

I had never realized just how vast the Pacific Ocean is until I sailed down the northeast trades. Three- to four-week passages under sail are commonplace. The Spanish explorer Ferdinand Magellan took ninety-eight days to sail from Patagonia to the Marianas in 1520–1521. Magellan was struck by the benevolence of the equatorial Pacific, where cool air sinking from higher latitudes toward the equator is pushed sidewards by the ubiquitous Coriolis Force to create the ever dependable northeast trade winds.

The west-blowing trades that carried me effortlessly to Bora Bora also push against the sea, blowing the warm surface water ever westward. It piles up in the far west, forming a pool of warmer ocean thousands of kilometers across. As the warm water moves west, colder water from the depths of the ocean flows to the surface near South America to take its place. The eastern Pacific is downright cold, even close inshore. Little moisture evaporates from it, so rain clouds rarely form. The Galápagos Islands and the Peruvian coast receive almost no rain year-round. The Baja Peninsula and the California coast have long dry seasons and even years of almost total drought.

Far away in the western Pacific, seas are much warmer. Moist air heated by the warm ocean rises, condenses, and forms massive rain clouds. Day after day, threatening clouds mass on the far horizon, sometimes rumbling with thunder or lit by distant lightning flashes. The heat and humidity climb to almost unbearable levels. Beast and human alike await the coming rain with tense anxiety. Then, one day, the clouds open with scattered showers, then a deluge of rain, as the monsoon bursts over Southeast Asia and Indonesia. Life-bringing moisture waters the fields and fills irrigation canals for another year.

The upward-rising warm air creates a vacuum, which draws in cooler air from the east to replace it. The northeast trades strengthen in response, pushing yet more warm water to the west. In a vast, self-perpetuating cycle, a two-way convection flows between South America and the southwestern Pacific, keeping the east dry and the west wet and allowing humble sailor folk like me to travel the great ocean like a highway.

When Jacob Bjerknes, the first scientist to identify the Walker Circulation that has governed Pacific Ocean weather patterns for thousands of years, published his famous hypothesis in 1969, he pointed out that each ENSO (El Niño–Southern Oscillation) episode has a unique personality: Some are intensely strong, others are weaker and shorter-lived. But all of them show remarkable similarities in their development, mature phases, and demise.

Only a narrow coterie of scientists took note of Bjerknes's theories until 1972–1973. In that year, catastrophic droughts gripped Central America, West Africa's Sahel, India, Australia, and China. Global fish catches declined significantly. The Soviet Union had such a disastrous harvest that it was forced to import large quantities of corn and wheat from the United States, with a resulting scarcity of grain on world markets. World food production per capita declined for the first time in more than two decades, raising concern about the ability of some nations to feed their citizens.

The new El Niño hit Peru especially hard. The anchovy fishing industry had boomed in the early 1960s. By the end of the decade, fish meal produced about one-third of the country's foreign earnings. Despite warnings from both local and international fisheries biologists, the government did nothing to curb overfishing. Now the fishery collapsed, setting off economic shock waves across the developed world. American poultry producers found themselves without fish meal. With expert advice, they turned to soymeal as a substitute. Soymeal prices rose higher than those for wheat, so North American farmers not unreasonably planted thousands of acres of soymeal for animal feed instead of wheat. Meanwhile, the global demand for wheat soared just as supplies were at their lowest level in decades, raising the real specter of a serious world food shortage.

The 1972–1973 El Niño served as a wake-up call for governments and scientists around the world, much as the 1899 famine had made the Indian government get serious about predicting monsoons. The threat of global hunger raised pressing questions about the threats posed by future El Niños and other large-scale climatic phenomena. Could one predict El Niños by tracking climatic anomalies in the Pacific and farther afield? Could long-range forecasts based on observation of these anomalies give governments the chance to prepare for excessive rainfall or severe drought? For the first time, meteorologists throughout the world began to see the Christmas Child as a climatic spoiler of the first magnitude.

Jacob Bjerknes and his predecessors achieved remarkable results with what we would now consider grossly inadequate tools. Their perceptive hypotheses have survived a quarter-century of intense research. A century after Alfonso Pezet introduced El Niño to an international audience, we can, at last, paint a reasonably convincing portrait of its birth, life, and death.

Today an army of scientists and an extraordinary array of satellites and scientific instrumentation record El Niño's every move. Truckloads of data make up today's El Niño portraits, much of it from decidedly low-tech sources. High-speed computers house enormous data banks of sea surface temperature records collected by merchant ships near the equator for over a century. Some land-based sea temperature observations from places like Puerto Chicama on the Peruvian coast date back to the 1930s. Daily observations of atmospheric pressure and rainfall flow in from hundreds of weather stations, just as they came to Sir Gilbert Walker's clerks in the 1920s, but now they are downloaded automatically into computers. Some locations, like Darwin, Australia, have records going back more than a century. The researchers collect fisheries records from California and South America, even written accounts of weather conditions by fifteenth-century Spanish colonists in Ecuador and Peru.

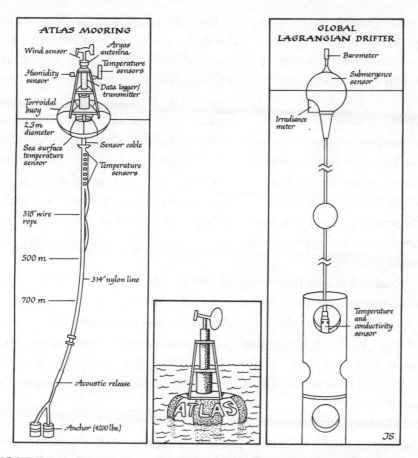

FIGURE 3.1 Five nations—the United States, France, Japan, Korea, and Taiwan—support the Tropical Atmosphere Ocean (TAO) array. This complex data-gathering system includes about seventy moored buoys that take measurements across the Pacific and transmit them ashore via satellite. TAO, designed specifically to improve detection and forecasting of El Niños, employs two kinds of buoys. ATLAS buoys (left) measure ocean temperatures, air temperatures, surface winds, and relative humidity. A 525-meter-long sensor cable clamped to a wire rope hangs below the buoy, weighted with a 1,900-kilogram (4,200-lb.) weight to measure deepwater temperatures.

The TAO array joins forces with a fleet of "global lagrangian drifters" (right), which drift free. They carry barometers as well as subsurface temperature and conductivity sensors. Since 1990, data from about 2,500 TAO buoys has flowed to a receiving station at the National Oceanic and Atmospheric Administration's (NOAA) Pacific Marine Laboratory in Seattle.

For all their reliance on these low-tech records, both scientists and government officials have realized the importance of enlisting high technology to monitor all manner of atmospheric, land-based, and ocean phenomena from year to year and over the long term. Over the past quarter-century, they have blanketed the world with monitoring devices in an effort to make El Niños and other large-scale atmosphere-ocean interactions forecastable. The instrumentation packages include a network of tidal gauges; drifting and moored buoys, many far offshore; and devices for measuring the amount of water vapor above the oceans, an important consideration in weather forecasting. Satellites transmit buoy data and measure sea level heights, for rising levels are a sign of climbing sea surface water temperatures.

Researchers on both sides of the Pacific and much farther afield are studying tree rings from ancient forests, underwater corals, even ice cores from Greenland and Andes ice sheets to search for evidence of great El Niños in the past. They seek answers to intriguing questions. How long have El Niños formed in the Pacific? Is there any reason to believe El Niños have lately become more frequent or more intense?

All this data flows into supercomputers in the United States and elsewhere, then into climate-prediction models used to generate long-term weather forecasts. The models are based on the fundamental laws of physics that govern motion, thermodynamics, and the behavior of water. Tens of thousands of lines of computer code describe the influence of factors such as mountains, clouds, and ocean temperatures. When ocean temperatures recorded by offshore buoys change dramatically because of developing El Niño conditions, the worldwide forecasts generated by these models differ radically from normal weather patterns.

The air in the anchorage off Huahine in French Polynesia is heavy and still, the water a dark, limpid gray. Sultry waves break on the beach by the town, bringing momentary flashes of white to the monotone. Tiny bugs swarm out from the land, buzzing, seeking human flesh. I sit sweating quietly under the awning, waiting for wind, thanking Providence that we are safe at anchor. I imagined slopping around in the open ocean, kilometers from anywhere, sails slapping in the lazy swell. Thick stratus clouds mantle the sky, pressing down on the land. The world waits for rain. Even the birds are silent, the palm trees still as statues. We perspire and wait for the rain to fall, the trades to resume.

The winds soon return, blowing away the depression and gloom. But we have a taste of what happens when the Walker Circulation's cycle falters. For some unknown reason, every few years something in the Pacific's perpetual motion machine hesitates. The ever-present northeast trades slacken and sometimes even die away completely. An El Niño episode is under way.

As the trade winds shift to idle speed, the inexorable forces of gravity kick in. Westerly winds now increase over the sea surface east of New Guinea, generating Kelvin waves, the internal waves below the ocean surface that push surface water across the tropical Pacific. The warm water that the trade winds had piled up in the western Pacific begins to flow backward to the east.

The pool of warm water in the western Pacific extends below the surface for some two hundred meters to a zone, known as the thermocline, where the sea temperature changes abruptly. Below are the much colder waters of the deep ocean. In the eastern Pacific, the thermocline is much closer to the surface, so sea temperatures are cooler. When the Kelvin waves push warmer water eastward, the thermocline sinks in the east, sometimes by as much as thirty meters. At the same time, with the trades no longer pushing water westward, sea levels off the Americas rise by several centimeters. Kelvin waves cause the same effect in the Pacific that you obtain by moving around in a bathtub. Satellite data show that a single Kelvin wave takes about ten weeks to travel across the Pacific basin—one-third of the earth's circumference, which works out to an average speed of about seven to eight kilometers an hour.

When the northeast trades blow, the warm ocean heats the overlying atmosphere in the western Pacific. Convection causes air to rise and rain-bearing clouds form. The cooler sea temperatures in the eastern Pacific inhibit cloud formation and rainfall in regions like the Peruvian coast. When the trades falter, the circulation pattern changes. The Kelvin waves created by the westerlies raise sea surface temperatures over much of the tropical Pacific, bringing thunderstorms into the central and eastern Pacific. Air pressure at sea level drops at Tahiti and rises at Darwin, Australia. Positive feedback develops between the ocean and the atmosphere, ultimately bringing on El Niño conditions.

The quieter trades alter wind patterns and currents throughout the tropics. The Walker Circulation reverses its flow. Rising warm water in the east sucks in air from the west. The trades may even change direction completely, giving sailors a rare fair wind from Fiji to Tahiti. Now the heated water is in the east, and the cooler water off Australia and Southeast Asia. Clear skies and little evaporation bring searing droughts to Australia and Indonesia while rain clouds form over the Galápagos Islands and the arid Peruvian coast. A hundred years' rain can fall in a few days.

In 1982–1983 the Galápagos Islands received 2,770 millimeters of rain, almost six times the normal amount. The number of flightless cormorants fell by 45 percent, while 78 percent of the rare Galápagos penguins perished. On some islands, 70 percent of the marine iguanas starved because red algae, nourished by the much warmer water, replaced the green algae that forms the lizards' staple diet. The normally barren islands blossomed with unfamiliar plant growth, which encouraged the spread of introduced pests such as fire ants and rats. After the 1983 event, most native species recovered, but recovery will be much harder in future years because the number of alien species increases each year.

Kelvin waves are hard to detect. They travel about thirty meters below the surface in what the oceanographer Nicholas Graham of the Scripps Institute calls "a weird sloshing effect." The only way scientists can detect these waves is with satellites that pick up the subtle undulations in sea level produced by their passage. Early in 1997, for example, the NASA oceanographer Anthony Busalacchi detected swarms of such subsurface waves heading across the Pacific toward the Peruvian coast. Eventually they hit the continental shelf, then branched north and south toward Chile and Alaska, carrying masses of warm water that brought a winter heat wave to Santiago, Chile, and tropical fish to San Francisco's Golden Gate.

With the sudden influx of heat in the eastern Pacific, the vast reservoir of warm, moist air over South America bulges dramatically, then cascades outward, disrupting the vast air flows that circle the earth. The jet streams lurch north, bringing heavy rains and severe storms to Mexico, California, and even the Pacific Northwest. Then the same stream crosses the Rocky Mountains, keeping arctic air from the far north out of the United States, so that Chicago and New York enjoy unusually mild, wet winters. Strong easterlies blow in the Caribbean, while displaced jet stream effects bring storms to the Gulf of Mexico and an abnormally cool winter to the South. Powerful east winds and winter storms also flow over Brazil, Chile, and Argentina.

Far away on the other side of the Pacific, Australian sheep farmers suffer through months of parching drought. Huge forest fires mantle much of Indonesia in suffocating smoke and grime, as they did in 1983 and 1997. Many times, but not always, a strong El Niño coincides with monsoon failure in India. Meanwhile, the Christmas Child creates havoc far downwind, eventually influencing weather in Africa, the North Atlantic, and even the eastern Mediterranean Sea.

Once the warmer waters have spread into the central and eastern tropical Pacific, surface temperatures begin to respond to changes in wind speed and direction. Some of the Kelvin waves rebound off South America. The west-going Rossby wave forms when the Kelvin wave bounces off the coast of South America, taking about nine months to flow west across the Pacific to Asia. The reflected waves eventually hit Asia again and rebound afresh. This time the double bounce has the opposite effect on the thermocline, making it once again deeper in the west and shallower in the east. The warm-water pool thickens in the western Pacific as the easterly trades strengthen

50

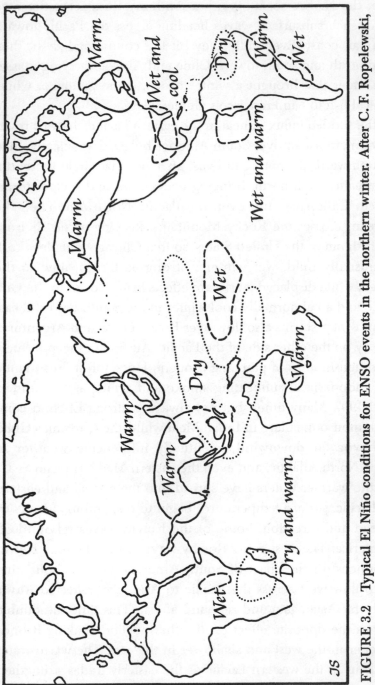

FIGURE 3.2 Typical El Niño conditions for ENSO events in the northern winter. After C. F. Ropelewski, "Predicting El Niño Events," *Nature* 356 (1992): 476–477; redrawn with permission from *Nature*; Copyright © 1992 Macmillan Magazines Ltd.

again. Upwelling renews, cooling the surface waters in the east, and El Niño becomes La Niña (Spanish: "young girl"), its cool opposite. We still know little about these "cold" events, which we often consider simply normal weather, as if El Niño with its many anomalies simply vanishes until a reincarnation appears years later. One would be hard put to define any point in these endless oscillations as "normal" Pacific weather. However, the unending cycle of change ensures that each El Niño contains the seeds of its own destruction.

El Niño is a chaotic pendulum, with protean mood swings that can last months, decades, even centuries or millennia. The pendulum never follows exactly the same path, for even minor variations in wind patterns can cause dramatic changes down the line. But there is an underlying rhythm to the swings, like a set of musical variations endlessly circling a central theme. Computer simulation after computer simulation produces the same general kinds of virtual El Niños, but none are ever identical. Nor do they occur at regular intervals. Tree-ring studies based on teak trunks from Java, fir trees in Mexico, and bristlecone pines in the American Southwest document years of higher rainfall, shown by thicker growth rings, about every 7.5 years until 1880, when the frequency increased to about every 4.9 years. La Niñas occurred every 10.0 years before 1880 but have since made their appearance at intervals of 4.2 years. From research on ocean corals and ice cores from mountain glaciers we know that ENSO has been a factor in world weather for at least five thousand years, and perhaps much longer.

The ENSO cycle is a powerful engine of global climatic patterns, swinging from one extreme to the other, from El Niño to La Niña and back again. Many scientists believe it ranks only behind summer, fall, winter, and spring as a cause of periodic climate change.

In 1935 a Swedish meteorologist, Anders Ångström, first used the word *teleconnection* to describe linkages between climatic anomalies in widely separated parts of the world. It has since become somewhat of a buzzword in the atmospheric sciences. Bjerknes's identification of the Walker Circulation heightened interest in the climatic linkages between ENSO and such disasters as droughts in Africa and floods in Chile and California.

Two German scientists, Hermann Flohn and Heribert Fleer, took the scientific bull by the horns in an article in the journal *Atmosphere* in 1974. They began in the equatorial Pacific and developed a chart showing possible teleconnections between climatic anomalies in tropical locations around the world. Flohn and Fleer started with warming sea temperatures in the Pacific and found these linked almost invariably to El Niños off South America, as well as to recurrent droughts in northeastern Brazil. The mature stages of El Niños affect sea surface temperatures so strongly that an area of high pressure (an anticyclone) can develop in the upper troposphere of the central Pacific, intensifying the subtropical jet streams on the north side of the anticyclone and thus bringing exceptionally heavy rain to southern Brazil, Paraguay, and northern Argentina, as well as floods and storms to the California coast and the Gulf of Mexico.

Rising sea surface temperatures in the equatorial Atlantic bring drought to West Africa. Water levels decline along the shores of Lake Chad, at the southern edge of the Sahara. Farther east, Nile River floodwaters at Aswan, Egypt, are lower than usual, and drought can affect India. Even farther east, persistent droughts in Australia often coincide with monsoon failure in India and have a close connection to the seesaw movements of the Southern Oscillation.

Flohn and Fleer's paper triggered a new era of research into the global effects of the Christmas Child—much of it based on both computer modeling and internationally supported ocean, land, and space-based observations—that leaves us in no doubt that ENSO is a major driving force in the global weather machine. For example, as sea surface temperatures peak during an El Niño event, temperatures over most of the tropics increase by about one degree. During cold events, air temperatures cool slightly. These swings are large enough to significantly affect average global temperatures. But to what extent are these temperature fluctuations caused by El Niño rather than other atmospheric changes? We do not know.

There is a clear link between El Niños and storms. As the Pacific trade winds weaken, tropical cyclones criss-cross French Polynesia with sometimes dramatic frequency. In December 1982, Cyclone Lisa hit the Tuamotus with 160-kilometer-per-hour winds. Another cyclone, Nano, lashed the Marquesas with winds of 225 kilometers per hour and nine-meter seas in January, bringing more than ten solid days of rain that triggered gigantic mudslides, destroyed roads and houses, and drowned most livestock. Soon afterward, Cyclone Reva, known locally as "the thrasher," passed within 120 kilometers of Papeete, Tahiti, uprooting trees or reducing them to stubble, peeling off roofs, and turning the harbor into a froth of white water. In Cook's Bay on Moorea Island, where the explorer James Cook anchored in 1769, savage gusts from Cyclone Veena knocked the yacht *Suka* on its beam ends while it lay at anchor. Said the skipper, Ray Jardine: "The boat righted and was quickly knocked down the other way, disgorging her port lockers. I crawled outside to bring in the dinghy, and found it flying behind the stern like a flag."[1] *Suka* dragged its anchor but fortunately came to rest only nine meters from a reef. At least nine tropical cyclones zigzagged across the central Pacific in 1997–1998.

Along the equator, a normal north-south circulation known as the Hadley Circulation connects the tropical atmosphere with that of northern latitudes. The Hadley carries winter storms northward toward Alaska. El Niño disrupts this pattern. As storm activity shifts eastward along the equator with warming tropical waters, strong westerlies begin to blow, and the storm track jogs eastward, hitting the California coast.

El Niño is a phenomenon of the tropics. For at least five thousand years, perhaps longer, it has exercised enormous power over millions of tropical foragers and subsistence farmers, as well as over great civilizations in river valleys, in rain forests, and high in the Andes Mountains. As we shall see, these societies have always been vulnerable to drought and flood. Today over 75 percent of the world's people live in the tropics, and two-thirds of them depend on agriculture for their livelihood. There are millions more inhabitants of this tropical world than there were even half a century ago. In West Africa's Sahel alone, the population has grown threefold since 1900. Vulnerability has increased everywhere as the carrying capacities of tropical environments have come under increasing stress.

Until recently, humans could only travel blindly on what the German statesman Otto von Bismarck once called the "stream of time," steering with more or less skill and experience from one ENSO episode to the next. No human society, however sophisticated or skilled at agriculture, could forecast the climatic rapids that threatened them. But now our computers and climatic models can predict El Niños well in advance, at least in a general way. In an overpopulated world, where millions of people live on the edge of starvation, this knowledge is of priceless economic, political, and social value—if we have the will to make the right decisions to avert disaster.

Although El Niño is the most thoroughly studied climatic phenomenon in the world, we still know little of its complex relationships to climatic changes in more temperate zones, where weather is controlled by alternating or clashing tropical and polar air masses. The Walker Circulation teaches us that atmosphere and ocean are vital elements in world climate. There is every reason to believe that both played important roles in the climatic regimens of the Ice Age as well. The geoscientist George Philander likens global climate to a performance by a giant orchestra: "Acting in concert, the ocean and the atmosphere are capable of music that neither can produce on its own. El Niño is an example of such music. An ever broader range of climatic fluctuations becomes possible once the interactions involve, not only the atmosphere and hydrosphere, but also the cryosphere [the Earth's ice volumes] and biosphere."[2] We have only just begun to understand the complex climatic forces that shaped the post–Ice Age world outside the tropics. As we shall see in the next chapter, some of them may lie deep below the North Atlantic Ocean.

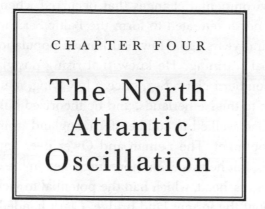

CHAPTER FOUR

The North Atlantic Oscillation

The wide loneliness of a great ocean breeds a humility that is good for the soul.

—*Peter Pye,* Red Mains'l

In 1931 a trawler working the North Sea's Leman and Ower Bank hauled up a lump of peat from a depth of thirty-six meters. The trawlermen cursed. Their nets routinely snagged and tore on waterlogged wood and mud lumps as they probed for bottom fish. Wearily, they bent over to throw the dark fragment overboard. The lump split open and a brown, barbed object fell on the deck with some peat still adhering to it. Fortunately for science, the skipper was intrigued and brought the find back to port with his catch.

The discovery came to the ears of a young Cambridge University archaeologist named Grahame Clark, a brilliant researcher who had just completed a doctoral dissertation on the Stone Age inhabitants of Britain at the end of the Great Ice Age. He identified the object as a finely made Stone Age deer-antler spearhead, identical to similar specimens found on both sides of the North Sea.

Clark had traveled widely in Scandinavia, where he studied the dramatic environmental changes that occurred when the great ice sheets of the north retreated to form the Baltic Sea, beginning some fifteen thousand years ago. How had human populations adapted to massive global warming? He knew that rising postglacial seas had flooded the southern North Sea, once a low-lying, marshy plain that joined Britain to the Netherlands, and he theorized boldly that Stone Age people had walked to southeastern England from what is now called the Continent. The Leman and Ower deer antler confirmed his hypothesis, but he was even more interested in the damp peat adhering to the spearhead, which had the potential to yield priceless information about the former land bridge. Clark handed over the precious artifact and its matrix to a young Cambridge botanist friend, Harry Godwin, one of the first scientists to specialize in fossil pollen analysis. Godwin and his wife Margaret assigned the antler-point peat to what pollen experts called the Boreal phase, when oak forests first spread northward into northern Europe in about 7500 B.C.[1]

The famous Leman and Ower spear point dramatizes the extraordinary changes in the North Atlantic world over the past fifteen thousand years. Fifteen millennia ago, huge ice sheets mantled North America from Alaska to Labrador and Greenland. Scandinavia lay under vast glaciers, as did the Alps and Pyrenees. Sea levels were at least ninety meters lower than today. The North Atlantic was a much colder ocean. Sea surface temperatures were cooler at the equator. Current circulations, ocean upwellings, atmospheric conditions, and weather patterns were probably different from those of today. Then massive global warming set in with breathtaking speed. The late Ice Age world vanished within a few millennia. Ice sheets shrank, sea levels rose rapidly, and summer and winter temperatures rose dramatically.

We still live in the midst of the Ice Age. At least eight major glacial cycles have enveloped the world over the past million years at intervals of about one hundred thousand years, with lesser surges each twenty-three thousand and forty-one thousand years. The last glacial cycle ended abruptly some fifteen thousand years ago. If humanly caused global warming does not intervene, we will undoubtedly enter another cold period, perhaps within the next twenty-three thousand years or so.

Short-term thinking means little to students of the Ice Age, who measure climatic change in millennia. We do not know what drives the procession of glacial episodes, but they are probably caused by slow cyclical changes in the eccentricity of earth's orbit and in the tilt and orientation of its spin axis. These shifts have altered patterns of evaporation and rainfall and the intensity of the passing seasons. The geochemist Wallace Broecker believes these seasonal changes caused the entire atmosphere-ocean system to flip suddenly from one mode during glacial episodes to an entirely different one during warmer periods. He argues that each flip of the "switch" changed ocean circulation profoundly, so that heat was carried around the world differently. In other words, Ice Age climatic patterns were very different from those of the past ten thousand years.

If Broecker is correct, then today's climatic mode results from what he calls the "Great Ocean Conveyor Belt." Giant, conveyorlike cells circulate water through the world's oceans. In the Atlantic, warm upper-level water flows northward until it reaches the vicinity of Greenland. Cooled by arctic air, the surface waters sink and form a current that flows enormous distances at great depths, to the South Atlantic and Antarctica, and from there into the Pacific and Indian Oceans. A southward movement of surface waters in the Indian and Pacific Oceans counters the northward flow of cold bottom water. In the Atlantic the northward counterflow is sucked along by the faster southward conveyor belt, which is fed by salt-dense water descending from the surface in northern seas.

The Atlantic conveyor circulation has enormous power, equivalent, Broecker says, to one hundred Amazon Rivers. Vast amounts of heat flow northward and rise into the arctic air masses over the North Atlantic. This heat transfer accounts for Europe's relatively warm,

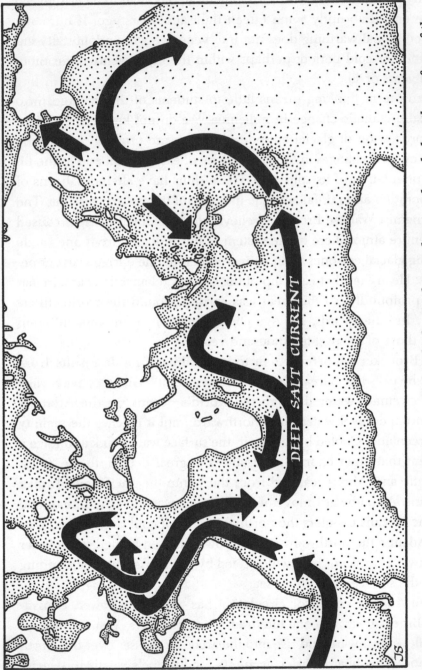

FIGURE 4.1 The Great Ocean Conveyor Belt, which circulates saltwater deep below the surface of the world's oceans. Salt downwelling in the North Atlantic Ocean plays a vital role in this circulation.

oceanic climate, which has persisted, with vicissitudes, through ten millennia.

We understand the Great Ocean Conveyor Belt only in the most general terms. The North Atlantic remains a mysterious confusion of swirling atmospheric streams, surface downwelling, and shifting ocean currents. Like El Niño, the chaotic equations of the atmosphere and ocean have left an indelible mark on human history thousands of kilometers from the open sea. Principal among these equations is the North Atlantic Oscillation.

From 1770 to 1778, the Danish missionary Hans Egede Saabye served on a remote station at Claushaven in Greenland, suffering through a series of intensely cold winters. In 1775 he wrote: "There has not, according to the oldest folk, been a winter like the past one in Greenland for the last thirty to forty years." The Greenlanders starved as the ice set in early and stayed very late. Many of them were "grey with hunger and so feeble that when at last there was open water, they were scarcely able to row their boats out."[2] Saabye had an interest in the weather and kept an eye on winters in his homeland. He was surprised to discover that "in Greenland, all winters are severe, yet they are not alike. The Danes have noticed that when the winter in Denmark was severe, as we perceive it, the winter in Greenland in its manner was mild, and conversely." Others also commented on this unexpected polarity of winter conditions when it would have been reasonable to expect bitterly cold weather in both regions. One hundred and twenty years later, scientists used forty-two years of monthly mean temperatures from Jakobshavn on Greenland's west coast and Vienna, Austria, to document what Saabye had suspected. They soon found a similar seesaw between Jakobshavn and Oslo, Norway.

In 1924 Sir Gilbert Walker used some of his statistical correlations to observe that there was a tendency for pressure in the North Atlantic to be low near Iceland in winter when it is high near the Azores and southwestern Europe. He called this the North Atlantic

Oscillation (NAO). Just as his Southern Oscillation in the Pacific reflected the difference in pressure between Darwin and Tahiti, Walker's North Atlantic Oscillation became an oscillating index that compares pressures between a persistent high over the Azores Islands and an equally prevalent low over Iceland. The atmospheric pressures of both centers fluctuate constantly, resulting in different pressure gradients between north and south. At the same time, when sea level is above normal in the Azores, it is below normal in Iceland, and vice versa. For many years scientists have known that the North Atlantic Oscillation is the dominant mode of atmospheric behavior in the North Atlantic region—especially in the winter months, when it can account for fully one-third of the total variance in sealevel pressure.

The North Atlantic Oscillation fluctuates constantly, but on a longer time scale than the year-by-year shifts in El Niños. A high index for the NAO indicates low pressure around Iceland and correspondingly high pressure off Portugal and the Azores, a condition that gives rise to strong westerly winds. During the winter, when the NAO is strongest, strong westerlies bring heat from the surface of the Atlantic to the heart of Europe, together with powerful, wet storms. The same westerlies keep winter temperatures mild, which makes British and northern European farmers happy. However, the same mild winters produce dry conditions in southern Europe. Portuguese and Spanish olive growers complained bitterly in the early 1990s when years of high indices decimated their harvests. Gstaad and other fashionable Alpine ski resorts suffered through unusually poor snow conditions.

In contrast, a low index means shallower pressure gradients, weaker westerlies, and much colder temperatures over the European continent. Cold air from the north and east flows from the North Pole and Siberia. Snow blankets Europe, and icy road conditions cause chain-reaction accidents and other havoc on German autobahns. Sometimes it even snows at Nice, on the French Riviera. However, Alpine skiers and tobogganers everywhere have a wonderful time.

The fluctuations of the North Atlantic Oscillation are the primary engine influencing weather on a year-by-year basis from eastern North America to Siberia, and from Greenland to the equator and perhaps beyond. David Parker of the Hadley Center for Climatic Prediction and Research in Bracknell, England, says that the NAO's winter oscillations account for about half the variability in monthly temperatures in central England.

The NAO fluctuates in decade-long patterns, and sometimes even longer periods. We have accurate records of long-term variability going back to 1864. A central Greenland ice core has shown that similar fluctuations extend back at least seven hundred years.

At its extremes, the NAO produces memorable weather. The turbulent 1880s brought extremely cold winters to Victorian England at a time of considerable social unrest. Hundreds died of cold in London's teeming slums. Between about 1900 and 1930, the Icelandic low was deeper, so strong westerlies kept temperatures mild over Europe, except from 1916 to 1919 when millions died in the climactic Western Front battles of World War I. My father served in the British infantry on the Somme. His diary contains entry after entry complaining of bitterly cold east winds that blew unrelentingly across the front line. For instance, on February 13, 1916, he recorded: "My birthday, remarkable for a welcome tot, but it hardly stopped the constant shivering. All quiet on the German side. I think we are all too cold to do anything but just survive." Four days later: "Private Williams froze to death before dawn. The idiot went to sleep on watch and never woke up. We had to pry his frozen hands off his rifle. Still the arctic wind blows, hour after hour."

A low index cycle twenty-five years later, in the 1940s, enveloped Europe in savage winter cold at the heart of World War II. Extremely cold east winds and blizzards brought subzero temperatures to the Russian front. Hitler's armies faltered as the oil in German panzers froze and thousands of hungry soldiers froze to death. The cold and the ferocity of Soviet resistance defeated the führer's invasion. In

1944 the Battle of the Bulge, in the forests and mountains of the Ardennes, began when German tanks attacked in the middle of a cold snap, hoping to catch the Allies off guard. They almost succeeded. Perhaps the worst winter was in 1947, when food and fuel rationing made the bone-chilling winds even harder to bear. That Christmas, the British cartoonist Osbert Lancaster penned a memorable cartoon of a poor man gathering wood under a full moon with a variation on the carol "Good King Wenceslas," lampooning the bureaucracy responsible for coal rationing:

> *Brightly shone the moon that night,*
> *'tho the frost was cruel.*
> *Extra brightly so's to spite*
> *the Minister of Fuel.*[3]

The 1950s were somewhat kinder, but the 1960s brought the coldest winters to Britain since the 1880s. I remember skating for kilometers on the frozen River Cam in Cambridgeshire a few days after having rowed on it in water so cold that spray froze on our oars. Over the past quarter-century the index has been high, bringing warmer winters again. These high indices, the most pronounced anomalies ever recorded, have made a great contribution to the observed warming of the Northern Hemisphere over the past twenty years. At the same time, the indices of recent years have triggered more winter storminess in the North Atlantic, higher wind speeds, and greater wave heights, including a gigantic wave that swept the *Queen Elizabeth 2* in midocean in 1996. The liner survived, but the waves smashed as high as its bridge. Increases in winter rainfall over Scandinavia have allowed the Norwegian government to sell surplus electricity generated from full hydroelectric reservoirs to other European countries. In recent years the growing season in Scandinavia has been as much as twenty days longer than normal. One can imagine the importance of these bonus days, during similar low-index cycles, to farmers thousands of years ago.

We live in a world of constant climatic transition, but the persistent warming of recent years has raised new worries. Have our depletion of the ozone layers and intensive use of fossil fuels started a process of humanly generated global warming, taking the world into strange climatic seas? No one knows for sure.

Every meteorologist agrees that the NAO is a primary factor in orchestrating hemispheric-scale climatic fluctuations over the North Atlantic Ocean and adjacent lands. While the workings of the Southern Oscillation and the Walker Circulation in the Pacific are relatively straightforward to reconstruct, at least in general terms, the NAO baffles easy explanation, partly because the atmosphere over the North Atlantic is perennially chaotic. Computer models make the point. The El Niño–Southern Oscillation (ENSO) cycle operates primarily as a result of changing ocean temperatures. Try to simulate its oscillations with a steady ocean temperature factored into the program and ENSO shuts down. Do the same with the NAO and it oscillates nicely. This strongly suggests that the ever changing atmosphere plays a major role in the unpredictable behavior of the North Atlantic Oscillation. Yet atmospheric conditions change by the hour and the day, whereas the North Atlantic Oscillation fluctuates over decades. Scientists who have modeled the NAO suspect that some other compelling force is also at work, probably in the depths of the Atlantic. They know that ENSO results from powerful interactions between the atmosphere and the ocean. Perhaps the NAO is the product of the same double act played out according to far more complex rules buried deep in ocean waters.

Researchers at the National Center for Atmospheric Research in Boulder, Colorado, and the Woods Hole Oceanographic Institution in Massachusetts have studied a series of years with similar high-index NAOs. They examined the NAO in the preceding fall and in the following spring, summer, and fall. They found that during the warmer parts of the year the NAO wandered aimlessly. But when winter came, the NAO appeared to revert to the state it was in during the previous year. Some oceanographers believe the ocean reminds the atmosphere what to do each year. What would be the likely candidates to give such reminders? One major player must be the ocean currents, notably the Gulf Stream and its relative, the North Atlantic Current, which form a critical component of Broecker's Great Ocean Conveyor Belt.

The warm waters of the Gulf Stream cascade northeastward up the Florida Strait to Cape Hatteras, pushing sailors northward in fine style. Five and a half centuries ago, pilots of Spanish treasure galleons used this current-driven highway to make their way north before heading eastward on the wings of the prevailing westerlies offshore to the Azores and home. The Spaniards learned from hard experience what the British Admiralty's *Ocean Passages for the World* proclaims to modern-day sailors bound for Europe from the Caribbean and Florida: "The great object for sailing vessels is to get North into the Westerly winds as speedily as possible."[4] So you ride the Gulf Stream. With a little luck and fair winds, you are home free.

Immediately north of Cape Hatteras, the Gulf Stream begins a gradual turn east into the Atlantic to pass south of the Georges Bank and the Nova Scotian Banks. The cold Labrador Current defines its northern edge quite clearly. East of about longitude forty-six degrees west, the Gulf Stream widens and weakens by fanning out along the east side of the Grand Banks. The warm water then cascades northeast and becomes the North Atlantic Current. Soon the current divides into northern and southern branches. The northern part moves toward the British Isles and the adjacent European coasts, passing off the northwest Scottish islands and along the Norwegian coast, with a branch moving toward Iceland and warming Greenland's southern tip. About latitude sixty-nine degrees north, above the Lofoten Islands, the warm water again bifurcates, with the western arm setting north toward Spitzbergen, where it enters the Arctic Basin.

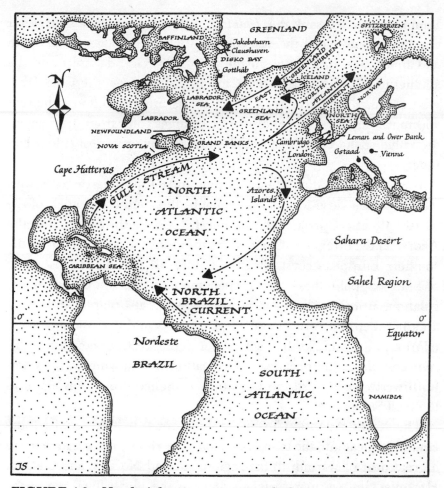

FIGURE 4.2 **North Atlantic currents and places mentioned in Chapter 4. These currents circulate surface water in the Atlantic, the North Atlantic Current bringing warm water to northern waters, where downwelling occurs.**

Here the current picture becomes more complicated. The East Greenland Current eddies down from the Arctic Basin around the Greenland coast into the Labrador Sea, where it joins other streams from the north and west to form the cold, south-flowing Labrador Current. This is the spoiler that brings icebergs east of the Grand Banks into transatlantic shipping routes. An iceberg carried by the

Labrador Current sank the *Titanic* in 1912. In winter the cold water continues south along the east coast of the United States as far south as Chesapeake Bay. Between the tail of the Grand Banks and the southern limit of the Labrador Current, sailors meet the Cold Wall, where ocean temperatures change by as much as eleven degrees within a few kilometers. The Atlantic changes color from the deep indigo blue of the Gulf Stream to the murky olive green of colder latitudes.

In the central Atlantic east of longitude forty-five degrees west (which passes close to western Iceland), the southern branch of the North Atlantic Current sets southeast, then southwest, joining the Azores and Portuguese Currents as they pass southward along the southern European coast. By latitude forty degrees north, just south of Madrid, Spain, these currents become part of a vast clockwise circulation, an oval cell of equatorial current that occupies the entire area between the African and Iberian coasts and the Atlantic coasts of West Indies and North America. Far to the west, the equatorial current completes the oceanwide circulation by carrying warm water northwestward toward the Caribbean, whence it emerges again as the Gulf Stream.

The North Atlantic Current is enormous. It helps redistribute equatorial heat into the temperate Northern Hemisphere and delivers a great deal of warmth to the regions around Norway and Iceland. However, the warmer surface waters of the north flush regularly, sinking toward the bottom, carrying atmospheric gases and excess salt. Two such major downwelling sites are known, one just north of Iceland, the other in the south Labrador Sea southwest of Greenland. At both these locations vast quantities of heavier, salt-laden water sink far below the surface. Deep subsurface currents then carry the salt southward on the Great Ocean Conveyor Belt.

As salt circulates southward from downwelling sites in the far North Atlantic, warmer water flows northward to compensate. So much salt sinks in the northern seas that a vast heat pump forms, caused by the constant inflow of warmer water, which heats the ocean as much as 30 percent beyond the heat provided by direct sunlight at these latitudes. What happens if the flushing fails? The pump slows down, the North Atlantic Current weakens, and temperatures fall rapidly in northwestern Europe. When downwelling resumes, the current accelerates and temperatures climb again.

Understanding the workings of the North Atlantic Oscillation will require years of long-term observations using ships, buoys, and computer models. The current suite of data and electronic projections shows how shifts in the NAO index coincide with changes in the Greenland, Labrador, and Sargasso Seas. Each region creates large volumes of homogeneous water, which affects the ocean over large areas. For instance, the warm Sargasso Sea of the central Atlantic is famous for its green weed. Christopher Columbus wrote in September 1492 that "it resembles stargrass, except that it has long stalks and shoots and is loaded with fruit like the mastic tree. Some of this weed looks like river grass, and the crew found a live crab in a patch of it."[5]

Columbus thought he was close to land when he was still well offshore. The Sargasso Sea produced vast quantities of warm water in the 1960s, when the NAO index was low. The Labrador Sea was quiet, but there was active downwelling in the Greenland Sea. In the early 1990s, the NAO index was very high. The Labrador Sea was very active, and everything was quiet in the Greenland and Sargasso Seas. Strong westerly winds cooled surface water over the Labrador Sea. At the same time, the north-flowing warm current warmed the Greenland Sea, made it less salty, and lessened convection. Flip the index, and the opposite occurs. Strong, dry, cold winds from the North Pole cool the Greenland Sea and increase convection. Meanwhile, the Labrador Sea misses the cooling winds, so downwelling slackens.

This circulation seesaw may be one of the triggers that change the North Atlantic Oscillation. As I write, the Labrador Sea is emerging from a period of vigorous downwelling in the early 1990s, which saw mild winters in Europe. In 1995–1996, the index changed abruptly from high to low, bringing a cold winter in its wake. Since then, the

signs are unclear. If we are witnessing a long-term shift to a lower index, then Europe can expect colder winters for a while. But the shifts
are very slow, for the sinking waters move only gradually southward.
The cold water generated in the Labrador Sea in the late 1980s and
early 1990s is only now flowing south under the Gulf Stream and
along the eastern edge of the North American continental shelf. In
time, colder surface temperatures could generate anomalies that will
wander across the Atlantic for many years.

Perhaps the NAO is another vast closed cycle, infinitely more
complex than ENSO, fueled by changes in its own output that flow
through the atmosphere and the ocean until they, in turn, cause the
system to flip to its opposite extreme. If such a closed cycle exists, it is
unlikely to operate in close harmony, simply because the time scales
of downwelling, sea level changes, and sea-surface temperature
anomalies rarely coincide. Even the frequency of the NAO's flip-
flops seems to vary considerably, from a period of one to two years a
century ago to a time scale of decades today.

Oceans and deserts are powerful engines in human affairs. The Sahara is another enormous pump, fueled by constant atmospheric
changes and global climate shifts. For tens of thousands of years its
arid wastes isolated the very first anatomically modern humans from
the rest of the world. But some 130,000 years ago, the Sahara received more rainfall than today. The desolate landscape supported
shallow lakes and semi-arid grasslands. The desert sucked in human
populations from the south, then pushed them out to the north and
west. The Saharan pump brought *Homo sapiens sapiens* into Europe
and Asia. And from there they had spread all over the globe by the
end of the Ice Age. But the pump shut down again. As glacial cold
descended on northern latitudes, the desert dried up once more,
forming a gigantic barrier between tropical Africa and the Mediterranean world. Fifteen thousand years ago, global warming brought
renewed rainfall to the Sahara. The pump came to life again. Foragers and then cattle herders flourished on the desert's open plains
and by huge shallow lakes, including a greatly enlarged Lake Chad.

Then, as the desert dried up after 6000 B.C., the pump closed again, with its last movements pushing its human populations out to the Sahel, where they live today. Like the North Atlantic Oscillation, the Sahara is a pump with the capacity to change human history.

Oceanographers in many countries keep a close eye on the downwelling sites in the Labrador and Greenland Seas. The flushing process oscillates constantly. It failed in the Labrador Sea in the 1970s, mounted a strong resurgence by 1990, and is now in decline again. Salt-sinking dropped by 80 percent in the Greenland Sea during the 1980s. Clearly, local failures are commonplace and may not have a great deal of impact. But what would happen if enormous quantities of freshwater suddenly flowed into the northern seas from melting ice sheets and glaciers, diluting salty surface waters before they became unstable enough to sink? Greenland is a huge freshwater ice sheet that could melt rapidly as a result of global warming. Theoretically, the same global warming would trigger rainfall in northern seas, dumping yet more freshwater into the North Atlantic. The abundance of freshwater would stop downwelling, halt the flow of warmer water to the far north, and cause an abrupt cooling in just a few years—with disastrous results for densely populated Europe and North America.

Perhaps this has already happened in the past. Between 9000 and 8000 B.C., after centuries of rapid warming, temperatures fell rapidly to near–Ice Age levels. This brief cold episode (described in more detail in Chapter 5) is a mysterious chapter in recent climatological history. Did it result from a sudden change in solar radiation, as some scientists believe? Or did a sudden influx of thawed freshwater from North American ice sheets cut off downwelling in the Labrador or Greenland Seas and halt the North Atlantic heat pump for more than ten centuries? The potential villains are in place—rapid global warming, cascades of melting freshwater, northward-flowing ocean currents, and salt downwelling. The Great Ocean Conveyor Belt may have flipped suddenly and triggered a millennium of renewed Ice Age climate.

The NAO and ENSO are two parts of a single, complex world climatic system. This climatic system oscillates on many time scales, confronting humanity with unusual and challenging weather at every season of the year. These oscillations—hot and cold, wet and dry—have always forced humans to adapt to rapid climatic change. For tens of thousands of years they did this by moving away from drought or flood or by using highly flexible social mechanisms that reinforced communal behavior so that no one starved. The rules of human existence changed after the Ice Age, about ten thousand years ago, when farming began: We settled in permanent villages and became anchored to our lands. Although we were losing the luxury of mobility, we still lived our lives within a context of enforced adaptation to repeated climate change. Civilizations rose and fell, great chiefdoms flourished in Europe and the Americas, and offshore sailors colonized the Pacific islands as human population growth accelerated everywhere. Our vulnerability to sudden climatic punches, so often accompanied by flood, drought, and starvation, has increased steadily over ten millennia, to the crisis point of today. Now there are over six billion of us, living in crowded cities, on increasingly exhausted farmland, and in marginal environments.

We still have much to learn about the twists and turns of atmospheric streams and ocean currents, the changes in sea-surface temperatures and convergence zones that fuel this remarkable engine of global climate change. We have always known that long-term climatic swings brought on major cultural and social changes in the past, such as the shift toward more intensive foraging with specialized tools directly after the Ice Age. Now a new generation of highly sophisticated research, in the depths of oceans, on ancient ice sheets, and in marshes and forests, has allowed archaeologists to observe how ancient societies wrestled with much shorter-term swings. For the first time we are on the way to understanding how the protean and unpredictable swings of global climate have affected the rise and fall of civilizations and many less complex societies for at least five thousand years.

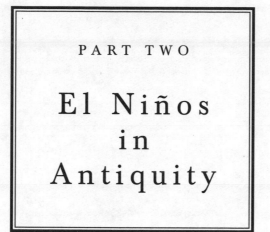

PART TWO

El Niños in Antiquity

... the Elysian Fields, where gold-haired Rhadamanthys
    waits,
where life glides on in immortal ease for mortal man;
no snow, no winter onslaught, never a downpour there
but night and day the Ocean River sends up breezes
singing winds of the West refreshing all mankind.
                    –*Homer*, Odyssey *IV*

CHAPTER FIVE

A Time of Warming

Behold, now, everything on earth,
Rejoiced afar at Ninurta, the king of the land,
The fields produced abundant grain,
The vineyards and orchard bore their fruit,
The harvest was heaped up in granaries and hills.
—Sumerian creation legend, Mesopotamia

Western Europe was a cold and windy place at the end of the Ice Age some fifteen thousand years ago. Where lush farmland and thick forests would flourish at the time of Christ, treeless, undulating plains stretched from the Atlantic coast deep into Eurasia far to the east. Low scrub hugged the ground, desiccated by bitterly cold winter winds that blew off the vast ice sheets that covered Scandinavia and the Baltic Sea. The climate fluctuated constantly, but in colder snaps winters would last nine months a year with subzero temperatures for weeks on end. Short, often hot summers brought relief to the sparse human population, few of whom wintered on the open plains. The largest forager groups lived in deep river valleys and sheltered areas in southern and southwestern Europe, where pine and birch forests grew and water meadows near fast-moving rivers supported game animals throughout the year. Their distant relatives in southwestern Asia exploited much drier landscapes, most living close to large rivers like the Euphrates, the Jordan, and the Nile. Surviving in these varied environments required ingenuity and opportunism, mobility and specialized technology in a seesaw world of climatic extremes and sudden shifts from near-modern temperatures to intense arctic cold. The Ice Age world was very dissimilar from our own, perhaps with radically different deep-ocean circulation patterns.

In about 13,000 B.C., a sudden roller coaster of global warming began. Glaciers shrank rapidly, releasing millions of tons of freshwater into the oceans. Dozens of familiar Ice Age animals large and small, like the woolly mammoth and giant ground sloths, became extinct within a few millennia. The warming shaped continents and enlarged oceans. Sea levels rose rapidly, flooding continental shelves and coastal plains and dramatically changing the world's geography. The Americas became a separate continent by 11,000 B.C., Britain an island five thousand years later. Rivers like the Mississippi and the Nile formed fertile floodplains as they filled up their deep valleys.

Pelting rains watered once-dry Southwest Asia and North Africa. Lush oak forests grew along the flanks of the Euphrates and Jordan river valleys in what is now Syria and Jordan. Shallow lakes and semi-arid grasslands covered hundreds of square kilometers of the Sahara. Only five thousand years after the Scandinavian ice sheet began retreating, forests covered most of Europe and sparse human populations had settled on the shores of the newly exposed Baltic Sea.

The very words "Great Ice Age" conjure up images of a world petrified in hundreds of thousands of years of profound deep freeze, when our skin-clad forebears hunted mammoth, reindeer, and other arctic animals. Deep-sea borings in the depths of the Pacific Ocean, coral growth series from tropical—and formerly tropical—waters, arctic and antarctic ice cores, and concentric growth rings from ancient tree

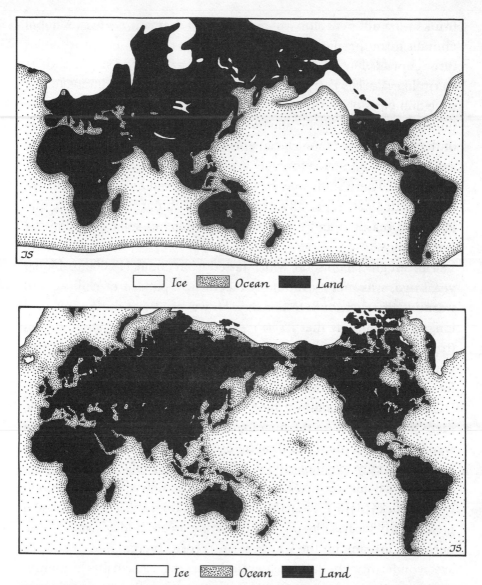

Ice ▨ Ocean ■ Land

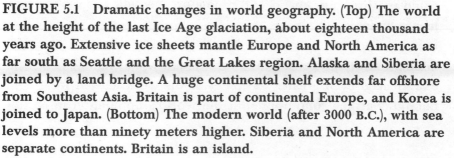

Ice ▨ Ocean ■ Land

FIGURE 5.1 Dramatic changes in world geography. (Top) The world at the height of the last Ice Age glaciation, about eighteen thousand years ago. Extensive ice sheets mantle Europe and North America as far south as Seattle and the Great Lakes region. Alaska and Siberia are joined by a land bridge. A huge continental shelf extends far offshore from Southeast Asia. Britain is part of continental Europe, and Korea is joined to Japan. (Bottom) The modern world (after 3000 B.C.), with sea levels more than ninety meters higher. Siberia and North America are separate continents. Britain is an island.

trunks tell a different story—of constant and dramatic swings in global climate over the past 730,000 years. Our remote ancestors lived through wild fluctuations from intense glacial cold to much shorter warm interglacials that sometimes brought tropical conditions to Europe and parts of North America. Over these hundreds of millennia, the earth has been in climatic transition for more than three-quarters of the time: At least nine glacial episodes have set a seesaw pattern of slow cooling and then extremely rapid warm-up after millennia of intense cold. These shifts may have been triggered by long-term astronomical changes, especially in the earth's orbit around the sun, that affect the seasonal and north-south variations of solar radiation the earth receives.

The last glacial episode, which lasted from about 118,000 to 15,000 years ago, was merely the most recent in a pattern of chilling and warming that continues to this day. However, it was during this prolonged cold episode that *Homo sapiens sapiens,* modern humanity, left tropical Africa and colonized Europe, Asia, and finally the Americas. For most of our time on earth, we modern humans have flourished in far colder, and often drier, environments than today's.

The dramatic global warming of the last fifteen thousand years is merely an incident on the ongoing climatic tapestry of the Great Ice Age. It is only a matter of time before the world cools down again—less than twenty-five thousand years, by one scientific estimate. However, this calculation takes no account of the new spoiler on the climatic block—fossil fuel–using industrial humanity. Human-caused global warming has already shrunk ozone levels. Today's world temperatures are among the warmest on record. Increasingly, scientists are wondering whether accelerated thawing caused by humanity could precipitate dramatic, and premature, climatic change within a few centuries.

Global warming is nothing new for humanity. Our remote forebears experienced even greater warmings as the world cycled from cold back to warmer conditions. But the world's population in earlier warm-ups numbered in the tens of thousands rather than the billions. Until the latest Ice Age, much of the world was still uninhabited. Human beings moved onto the Central Russian Plains less than twenty-five thousand years ago, and into the Americas, at the earliest, about twenty thousand years ago, probably later. There was plenty of space for everyone. Fifteen thousand years ago, most Stone Age people lived in tiny family bands and occupied home territories extensive enough for them to be able to move around freely, even over large distances, using highly flexible survival strategies. If a small African foraging band experienced two consecutive drought years, they simply moved into better-watered areas or fell back on less desirable plant foods, perhaps species that required more energy to harvest.

Our Stone Age ancestors were opportunistic, accustomed to sudden climatic change, and able to adapt to it in ways that became impossible when populations rose rapidly after the latest Ice Age. Until about eleven thousand years ago, when farming appeared, earth's population had not yet reached the critical point where it exceeded the natural carrying capacity of the land. Experience, low population densities, and the sheer flexibility of human existence allowed Stone Age foragers all over the world to ride with the punches of the global weather machine.

That world is gone. Hunting and gathering as a way of life began to disappear with the advent of agriculture. Africa is one of the few continents where you can still glimpse the ancient lifeway. Even there, the foragers have been in contact with farmers, herders, and Europeans for generations. The few surviving hunter-gatherers live in semi-arid environments or in rain forests; though they stay in regular contact with the wider world, they maintain a semblance of a lifestyle that has deep roots in the Stone Age. The !Kung San[1] of Botswana's Kalahari Desert ride out droughts using strategies developed thousands of years ago to combat serious food shortages. Similar strategies must have been used by Stone Age foragers throughout the Ice Age world.

The !Kung flourish in an arid environment where they live off hares and other small creatures, but above all off plant foods. To all intents and purposes, the !Kung are self-sufficient hunter-gatherers with reasonably adequate water supplies for six to eight weeks a year and more limited, permanent water holes to use the rest of the time.

Years of plenty sometimes give way to El Niño–induced droughts. The Kalahari is never well watered, so the !Kung are used to long dry spells, during which they fall back on the most reliable water holes and eat a far wider range of plant foods. Having consumed just a handful of them in normal years, they have a cushion of alternative foods to fall back on. Life is not necessarily comfortable, but at least there is food. At the same time, the !Kung resort to social obligations to distribute risk over as wide an area, and as many people and bands, as possible. Each family creates ties with others in a system of mutual reciprocity called *hxaro*.

Hxaro involves a balanced, continual exchange of gifts between individuals that gives both parties access to each other's resources in times of need. *Hxaro* relationships create strong ties of friendship and commitment.

Hxaro distributes risk by giving each party an alternative residence, sometimes up to fifty to two hundred kilometers away. Each family has options when famine threatens. At no time does an entire population uproot itself and descend on an area where food is available. During a drought, a high proportion of families leave to visit *hxaro* partners or relatives, relieving the stress on those who stay behind. Archaeologists believe similar mechanisms were in wide use elsewhere in southern Africa as early as 2500 B.C.

The *hxaro* strategies work well in egalitarian societies, where kin ties and trade partners are the key to risk management and survival. Such coping mechanisms, operating at the family level, spread risk over society as a whole as long as humanity does not overexploit the land. Social obligations and kinship, carefully acquired environmental information, and long foraging experience provide a simple safety net that helps people survive without exceeding the carrying capacity of the land. *Hxaro* is a relic of a long-vanished world where flexible lifeways, mobility, and low population densities allowed foragers to adjust easily to droughts, floods, and other caprices of ancient climate.

In 13,000 B.C., the world's hunter-gatherer population was approaching eight and a half million. For tens of thousands of years, the growth rate had been roughly 0.0015 percent per year as our remote ancestors expanded into deserts, tropical forests, and arctic regions.[2] Before fifteen thousand years ago, the ability of the world's environments to support animals and people still exceeded the needs of the human population. But late–Ice Age people had developed increasingly efficient ways of exploiting food resources of every kind. Now their numbers rose much faster, especially in food-rich areas. They, and their ancestors, had an encyclopedic knowledge of plant foods. I once showed a !Kung man a collection of five-thousand-year-old seeds from an ancient camp in central Africa. He not only identified every one but knew the food and medicinal value of each.

As worldwide global warming began and the glaciers retreated, the growing human population approached the limits of the world's ability to support foragers. No one had as much room to move around, and there were more mouths to feed. Unlike animals, however, humans can get around environmental limits by using labor, social relationships, and technology to acquire more food. As the glaciers retreated and the world became full, they did just that. Foragers everywhere adopted new and even more intensive food-gathering strategies. They specialized in plant foods, exploited caribou migrations, and harvested seasonal salmon runs. In favored areas like parts of southwestern Asia, Europe, and the Americas, local populations continued to increase steadily as people developed ever more efficient bone, stone, and wood technologies to exploit their environments.

It may seem incongruous to say "the world was full" when its population was that of one good-sized modern city. But few environ-

80

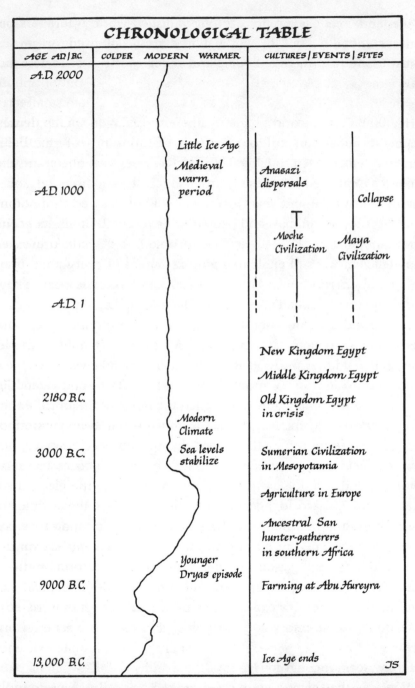

FIGURE 5.2 Chronological table of the major cultures and historical events in this book.

ments on earth could support more than one or two foragers per square kilometer. Local populations faced starvation unless they adjusted to the new realities of more circumscribed territories, less mobility, and more mouths to feed. People began to run out of options for making a living. Many hunter-gatherers crowded into small territories and permanent camps by rivers and lakes. Some of the densest populations flourished in food-rich areas like the shores of the Baltic Sea in Scandinavia and Egypt's Nile Valley, where inhabitants exploited fish, game, and plants from several environmental zones within easy reach. Danish archaeologists have recovered entire Stone Age camp sites preserved under the rising waters of the Baltic, complete with fish traps, canoes, and numerous perfectly preserved wooden artifacts such as barbed spears. The inhabitants could draw on so many food sources that they occupied the same locations for months on end.

Nevertheless, even the most prosperous of these Stone Age groups were far more vulnerable to sudden climate change than their predecessors had been, simply because they could not move. Every year brought a challenge of contrasts: hunger or plenty, good rainfall or drought, sudden floods or greatly reduced game populations. At the same time, people lived close to one another for much longer periods of time, even if they did not get on well together. Whereas once they had simply moved away, now they needed firm leadership, based on more than just long experience, that took account of public opinion and ensured an equitable distribution of food supplies in times of need. Crowded into somewhat larger communities than foraging camps, people now also needed leaders to mediate disputes over domestic matters and foraging rights. Societies changed profoundly as population densities rose and people lived closer to the edge of survival in a more crowded world. Because so many societies were close to exceeding the carrying capacities of their environments, short-term climatic change became an important factor in human existence for the first time.

We have long known that long-term climatic change profoundly influenced human evolution and history. Only recently have we also realized that short-term changes like El Niños or decadal swings of the North Atlantic Oscillation have had the power to change history and topple civilizations.

Our record of the Ice Age, with its gigantic climate swings, is incomplete, measured in millennia. This veiled mirror obscures all but the major fluctuations of global climate, so we know nothing of annual or decadal shifts in temperature or rainfall patterns. Nor do we know much about the complex atmospheric and ocean interactions that drove Ice Age weather. About fifteen thousand years ago, the mirror begins to clear. We can measure cold and warm snaps in briefer snippets of geological time, sometimes even centuries. And we can see how, with so many people now living on earth, these short, often sharp fluctuations helped change history, turning foragers into farmers and villagers into city dwellers.

After 3000 B.C., the mirror clears still more, and we can begin to study climate changes in decades and years, whether cycles of intense drought that fractured kingdoms or one-hundred-year rains in one season that caused powerful empires to collapse and thousands of people to starve.

People have always lived with occasional hunger and malnutrition, disease, and parasites. These hazards have been part of daily life and are something quite different from not being able to forage enough food to feed everyone living in one's territory year after year. The carrying capacity of the land is a delicate balance between population density and the ability of any particular environment to support that population indefinitely. This equation of people and sustainability allows for such variables as seasonal rainfall, droughts, and short-term climatic shifts like El Niños. The crisis comes when sudden climate change or population growth renders sustainability impossible.

Eleven thousand years ago, the fertile valley of the Euphrates River supported dense stands of wild cereal grasses and thick oak forests. Enormous gazelle herds migrated through the valley in spring and fall. Fish abounded in the slow-running river. Hundreds of forager families lived on the edge of the bountiful valley, where food was so abundant that they could live in one spot for most of the year. One such settlement was Abu Hureyra, a tiny hamlet of circular houses dug partly into the ground.[3]

Abu Hureyra stood close to the river floodplain and in close proximity to lush forests. Each fall the women of the village picked ripe hackberry and wild plums. They harvested thousands of ripe nuts from laden trees. The autumn nut harvest and dried gazelle meat provided staples for the cold winter months. There must have been lean years and hungry months, but as long as the rains fell, there was enough food to feed even a growing village population.

But the global warming that sustained this environment was interrupted.

Far to the north, the great Laurentide ice sheet that had covered eastern North America for 100,000 years was in full retreat. Enormous volumes of freshwater flowed into the North Atlantic. Armadas of icebergs broke off from the eastern margins of the ice cap and melted offshore. As much as half a meter of small rocks and debris from the Hudson Bay region dropped onto the floor of the Labrador Sea. The ocean was choked with sea ice, which reduced salt levels dramatically.

As the Laurentide ice sheet melted, Lake Agassiz, a huge freshwater lake, formed in the depression left by the retreating ice cap. At first, the meltwater in the lake spilled over a natural rock sill into the Mississippi watershed and flowed into the Gulf of Mexico. Twelve thousand years ago, the shrinking ice front opened up a new channel to the east. Lake Agassiz dropped rapidly as floods cascaded across southern Canada and into what is now the Saint Lawrence Valley. An enormous surge of freshwater washed into the already much-diluted Labrador Sea. Within a millennium, the downwelling that carried salt into the deep ocean and on its long voyage southward stopped altogether. The warm conveyor belt that had nourished global warming for three thousand years abruptly shut down. Far below the ice-strewn surface of the Labrador Sea, salt ceased to flow away from the northern ocean. Global warming stopped within a few generations. Bitterly cold north winds blew southward from the North Pole. The northern glaciers advanced once more, plunging the world into a millennium-long cold snap. European forests retreated in the face of the cold. Glaciers advanced in the Andes and New Zealand.

The Ice Age returned for a thousand years. Then, for reasons we do not understand, Atlantic downwelling suddenly resumed, and the conveyor belt switched on. Ten thousand years of warmer conditions began and continue to this day.

This event, called the Younger Dryas (after the polar wildflower *Dryas octopetala*), brought severe drought to the Euphrates Valley. A surge of cold from the north, bone-chilling winds, reduced rainfall, and much depleted nut harvests changed Abu Hureyra's world within a few generations. The nut-rich forests retreated almost one hundred kilometers away from the village. The people turned to wild cereal grasses and drought-resistant clovers to supplement the now-meager nut harvests. Even valley bottom plants became rarer, because the Euphrates only occasionally overflowed its banks. A thousand years earlier, the foragers could have dispersed into smaller groups and relied on ties of kin to see them through hungry years. But Abu Hureyra and its neighbors now housed many more families than the suddenly much drier valley could support. Local game populations crashed. Humans survived because they put their botanical expertise and food-gathering technology to work. Well aware that seeds germinate in the ground, they deliberately planted einkorn and wheat to supplement dwindling wild stands and to increase food supplies.

The supplements soon became staples. Within a few generations, the foragers became full-time farmers.[4] Abu Hureyra grew rapidly into a compact village of mud-brick houses separated by narrow alleyways, remarkably like many farming settlements in the region as recently as the nineteenth century. Now wheat and barley fields lapped the edges of the settlement, and there was just enough to eat in good rainfall years. At about the same time, other drought-stricken Southwest Asian communities domesticated wild cattle, sheep, and goats.

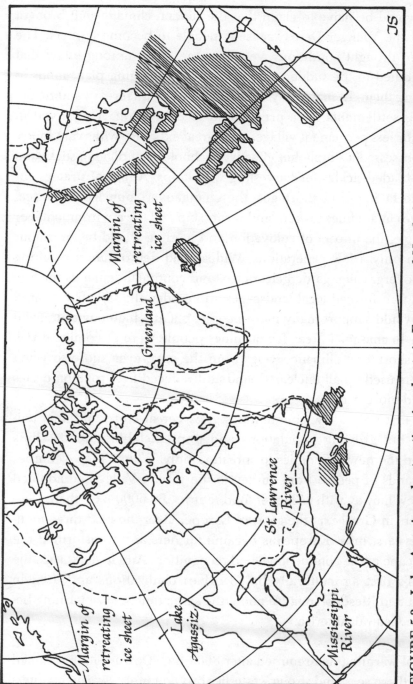
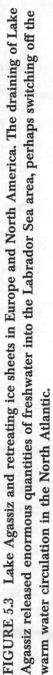

FIGURE 5.3 Lake Agassiz and retreating ice sheets in Europe and North America. The draining of Lake Agassiz released enormous quantities of freshwater into the Labrador Sea area, perhaps switching off the warm water circulation in the North Atlantic.

It would be naive to claim that the sudden climatic shift brought on by the Younger Dryas "caused" people to become farmers. The sudden drought conditions were but one factor among many that contributed to the sudden changeover from foraging plant foods to growing them. In many ways, the people of Abu Hureyra and neighboring settlements were preadapted to cultivation. They lived in more or less permanent villages, had great expertise with plant foods, and possessed a simple but effective technology of digging sticks and stone-bladed sickles for harvesting wild crops. A logical strategy for people facing food shortages, the changeover soon led to radical changes in attitudes toward land ownership and the environment. For example, the impact of cultivation and herding would be enormous within only a few generations. Widespread deforestation resulting from overgrazing, garden clearance, and continual demands for firewood transformed local landscapes and eroded the soil. The cleared land would support many more people, but at a high environmental price. In marginal areas for farming, people were at continual risk from short-term climatic swings from the start, even more so when farmers used up all uncleared land and reduced field rotation cycles to feed more people.

At first only a few communities grew cereal crops and kept animals, but growing populations and unpredictable climatic shifts caused the new economies to spread rapidly. Within a few centuries thousands of people from Turkey to the Jordan Valley had adopted the new lifeway with its higher food returns. By 6000 B.C., there were farmers in Greece, at the gates of Europe. Over the next three thousand years, most Europeans became farmers and transformed the landscape of the continent beyond recognition. And in 3000 B.C., the world's first agriculturally based urban civilizations appeared in Egypt and Mesopotamia. At about the same time, maize farming began in Central America.

Global warming had returned after 8000 B.C. Downwelling resumed in northern seas, and strong westerlies brought moist, warmer winters to Europe as the northern glaciers retreated into Norway's mountains and melted into the newly formed Baltic Sea. By 6000 B.C., dense oak forests stretched from the Atlantic coast to western Russia, from the Alps to southern Scandinavia. Europe's long summers were several degrees warmer than today's, and the seas rose rapidly to near-modern levels.

During these warming millennia, Europe was home to sparse populations of hunters and foragers, many living along sea shores, in river valleys, or by freshwater lakes. Most Europeans were constantly on the move, following the seasons of plant foods, exploiting salmon runs, pursuing migrating game. People lived in small bands of a few families that changed almost daily. A man and his son would die in a hunting accident, and his wife and younger children would join another band. Two brothers would quarrel, and one would simply move away to another camp. Everyone lived in a state of constant social tension and flux, but there was nearly always enough to eat, even in the coldest winters, because the people knew of so many different foods and could fall back on less palatable game or plants in hard times.

In about 6200 B.C., four centuries of cooler, much drier, near–Younger Dryas conditions sent a wave of drought across southeastern Europe and the eastern Mediterranean. Many farming settlements were abandoned as many communities retreated to the shores of permanent lakes and rivers, including the margins of the Black Sea, at the time a huge freshwater lake. Better rainfall and warmer conditions returned in about 5800 B.C. Complex detective work with deep-sea cores and pollen analysis has documented the environmental catastrophe that followed. The Mediterranean continued to rise, lapping at the crest of a natural earthen barrier at the mouth of the Bosporus Valley. The Black Sea lake lay 150 meters below the berm. Suddenly, around 5500 B.C., seawater washed over the rim. A gentle stream soon became a powerful torrent as cascading seawater eroded away the barrier in a horrendous waterfall that cut deep into the underlying bedrock. The lake, now part of the Mediterranean, rose rapidly, flooding low-lying shores, killing millions of fish, and drowning ancient farmlands near the shore. Thousands of farmers hastily moved far inland, up river valleys and onto higher ground, taking their customs, language, and farming culture with them.[5]

Many of these fugitives must have entered the forested European world. They lived by fire and ax, cultivating rich, weathered glacial soils, clearing the now-temperate continent of its boreal forests. Slash-and-burn agriculture is a wasteful way of farming land. Clear some fields, cultivate them for a short while, then move on when the soil is exhausted or when there are more mouths to feed. The first such farmers settled in the Danube Valley in about 5300 B.C. Within seven centuries their descendants had slashed and burned their way from the Balkans to the Netherlands and east into the Ukraine. Each planting season, thick gray smoke mingled with leaping flames in the hazy blue sky. The farmers prepared for this moment for days, felling small trees and low branches, spreading dry undergrowth over the ground between the massive oaks. The woodsmen used stone axes to topple the long, straight trunks, then hammered wedges into them to split long planks for their communal houses and barns. The flames moved fast, charring ancient oaks, thinning centuries-old growth and canopy to let in the sun. Once the gray and white ash cooled, the villagers spread it over the new fields and planted wheat and barley on the fertile ground.

Primordial oak forest gave way to an increasingly organized and intensively cultivated landscape over much of Europe within a few centuries. The farmers settled in homesteads where extended families dwelt in split-log, thatched longhouses, set among fenced fields and animal pens, separated from their neighbors by abandoned fields and the remnants of ancient forest. At first there was plenty of land to go around. But as farming populations rose, the gaps between individual settlements filled in, anchoring people ever more closely to their lands. Between years of plenty, the farmers sometimes suffered through unpredictable cycles of bitterly cold winters when they quietly starved. Over the centuries they developed technologies to in-

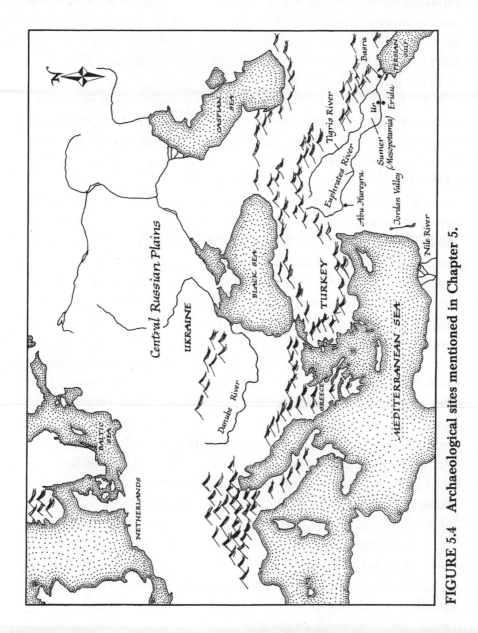

FIGURE 5.4 Archaeological sites mentioned in Chapter 5.

crease crop yields and create greater food surpluses—simple ox-drawn plows to turn over heavier soils, and large-capacity grain storage bins. They rotated the crops and their fields and used cattle manure to fertilize the soil. But there was no moving away from their increasingly crowded lands.

We humans became farmers because we had to. We changed the way we lived and interacted with one another almost overnight because we were at the mercy of distant Atlantic currents that brought rain and mild winters to our homelands. We have been at the mercy of short-term climate change ever since.

So far, luck has been with us. The world has enjoyed ten millennia of warmer climate since the end of the Younger Dryas. This was the last of the longer-term shifts to challenge humankind, and it lasted only a millennium, a blink of an eye by geological time standards. Since then, irregular shifts have caused centuries of unusual warming, like the so-called Medieval warm period of A.D. 900–1300, when vineyards flourished in northern England and the Norse colonized Greenland. A century later the somewhat grandiosely named Little Ice Age brought Europe significantly colder conditions that endured until the 1850s. The cold caused great suffering and brought the potato into fashion as a staple for Europe's poor. These centuries-long cycles were like irregular heartbeats in the steady rhythm of global weather. They caused much comment at the time but are mere background noise to the passage of climatic history. The most dramatic shifts have been on an even shorter compass—the decadal flips of the North Atlantic Oscillation, the constant seesaws of ENSO in the Pacific. These shifts can bring violent, unpredictable weather and threaten catastrophe, especially for societies living at the edge of survival, in marginal lands, or at high densities in exceptionally fertile environments. For the past five thousand years, ENSO and the NAO have killed millions of people, caused civilizations to collapse, and taxed human ingenuity to the limit.

The millennium 3000 B.C. was a major turning point in human history. The world's sea levels had stabilized at near-modern levels, and farming was well established throughout the Old World, except in game-rich tropical Africa and Australia. Global populations were rising faster than ever before. The world's first urban civilizations appeared in Egypt and Mesopotamia. This was the moment when short-term climatic shifts became important in human affairs. Dense urban populations in circumscribed environments require highly productive agriculture just to survive, let alone make it through drought cycles or natural disasters. For over five thousand years the household and the village had been the basic economic units that fed humanity. Families still fed themselves. Throughout history, the anonymous farmer always supported the elaborate superstructure of preindustrial civilization. Increasingly, agriculture became a matter of irrigation systems and canals, large-scale communal projects, taxation, and centralized grain storage. Governments and supreme rulers now became farmers, for they commanded the labor to build the irrigation systems, divert rivers, and transport grain. They used muscle power and technology to tame floodplains and produce three crops a year instead of one. In so doing, they placed many of their economic eggs in single baskets, a strategy that worked well until El Niños or other climatic anomalies brought droughts or torrential rains that swept away in a few days the work of generations.

In 2550 B.C., a Sumerian scribe in Mesopotamia complained of great hardship:

> *Famine was severe, nothing was produced*
> *. . . The water rose not high*
> *The fields are not watered*
> *. . . In all the lands there was no vegetation,*
> *Only weeds grow.*[6]

From his temple writing room in the city of Ur in southern Iraq, he would have looked out over a checkerboard of irrigation canals and densely packed fields. Normally, the crops stood green in their furrows, the ever present wind ruffling the blue waters of the canals. This year was different. Barley and wheat withered yellow in the parched fields. The canals were muddy rivulets, abandoned by the nearby Euphrates flowing sluggishly far below its banks. Farming had never been easy in this harsh land of sudden floods and seasons of drought. But now the droughts persisted year after year, while the people complained of draconian taxation and poor rations.

The Sumerians created the world's first civilization in Mesopotamia at a critical moment in history—when sea levels stabilized and short-term drought cycles related to the Southern Oscillation and periodic monsoon failures became a reality of life in Egypt and southern Iraq.

The Persian Gulf is probably the most studied stretch of water on earth. Nowhere deeper than one hundred meters, the seabed flattens into a gentle basin about forty meters below modern sea level. During the height of the last Ice Age, this was dry land. Even as late as twelve thousand years ago, rising ocean waters were only just entering the shallow basin. A deep canyon, which geologists call the Ur-Schatt River, carried the waters of the Euphrates and Tigris down to the much lower Indian Ocean. The cradle of Mesopotamian civilization did not exist. Drowned sand dunes under the northern Gulf and oxygen isotope readings from deep-sea cores tell us the entire region was very dry indeed.

After the Younger Dryas ended in about 8000 B.C., seas rose rapidly throughout the world. The rising Indian Ocean penetrated so rapidly into the Persian Gulf in some places that the water advanced northward eleven meters a year, forming a deep indentation in the desert. Fossil tree pollens from Lake Zeribar in Iran tell us that considerably more rain fell in the region at the time. By 6500 B.C., the rising waters had flooded the Ur-Schatt river valley and reached the present-day northern Gulf. A large marine estuary formed where the lower Euphrates flows today. When the sea stopped rising in about 5000 B.C., having flooded the site of modern-day Basra at the head of the Gulf and much of extreme southern Iraq, the northern portion of the Gulf was an enormous estuary fed by the Euphrates and Tigris. As sea levels stabilized, the delta filled in with silt, but the water table remained high.

Unfortunately, thick layers of river silt and rising sea levels put the Gulf's ancient shores and the banks of the Ur-Schatt beyond the reach of the archaeologist's spade. However, there may have been settlements like those at Abu Hureyra far up the Euphrates, where Stone Age foraging bands camped in desirable locations; they especially favored the margins of several ecological zones such as those on the Gulf coast, where the greatest diversity of food resources could be found. The harsh drought and cold of the Younger Dryas may have triggered a similar shift over to farming in this region as well. We will never know.

For the fifteen hundred years after 7000 B.C., tree and grass pollen profiles from southwestern Asian lakes show that southern Mesopotamia enjoyed an unusually favorable climate, with greater and more reliable rainfall than today. Farming communities prospered throughout the low-lying delta area by 6000 B.C. As estuaries formed ever farther inland and the river floodplains expanded after 5500 B.C., conditions became steadily drier. Rainfall faltered, and people once again turned to human labor and technology to boost the carrying capacity of their lands. The delta farmers developed irrigation into a fine art. Crop yields and population densities climbed rapidly at a time when the Gulf was moving inland at a remarkable rate. Communities near the water's edge had to move several times within a generation. In a tightly packed farming landscape, even a minor move altered jealously guarded territorial boundaries and sent political and social ripples through a wide area. We can imagine such an incident. Two families have cultivated fields near a stream for generations. Now the water course has vanished under rising seawater. Both households claim new land close to the new shore, but there is not enough well-watered soil to go around. Shouting erupts, then fistfights. A kin leader intervenes and settles the dispute. No one is happy with his decision to divide the land, but there are no other options, except to build new irrigation canals farther inland—but other neighboring villages want the same site.

The farmers of the day lived in a state of continual negotiation, in a volatile world where only strong leadership kept the peace.

Intelligent use of water supplies and irrigation technology paid handsome dividends and created ample food surpluses to support a growing population of nonfarmers: artisans, priests, and other officials. Village kin leaders adjudicated disputes and organized canal works. Centuries later their powerful descendants became the spiritual and political leaders of communities of much greater complexity. They were thought to possess supernatural powers and to be capable of interceding with the gods, who controlled the hostile forces of nature: the chilling winter north winds, the spring floods that carried everything before them, the choking heat and dust of summer. A leader's authority depended on his ability to intercede successfully—to see to it that the gods kept his subjects safe and well fed.

Within a few centuries agglomerations of villages became small towns, each dependent on its neighbors, linking the entire region in close trading networks. By 3000 B.C., the sea had stopped rising. River silt filled large estuaries as the climate became progressively drier. The low-lying landscape was a mosaic of desert, semi-arid plains, and lush estuaries and riverbanks. The towns became cities, several of which became the center of the world's earliest civilization.

Sumerian civilization was a patchwork of densely populated city-states competing ferociously for political and spiritual advantage. Theirs was an environment of violent contrasts governed by flood and irregular rainfall: A sudden drought cycle could topple a king. The most powerful of these leaders presided over cities like Eridu and Uruk, where great temples celebrated the all-powerful gods.

Eridu was the earliest city, the dwelling place of Enki, God of the Abyss, the foundation of human wisdom. "All lands were the sea, then Eridu was made," proclaims a much later creation legend. Enki had built his shrine on the shores of the primordial sea, which preceded creation. His word created order from the chaos of the waters and the mosaic landscape of the Sumerian world. As early as 4900 B.C., a small mud-brick shrine arose at the heart of a growing village. By 4000 B.C., Eridu covered about twelve hectares. A thousand years later, as many as six thousand people lived there in crowded neighborhoods of small mud-brick houses separated by narrow alleyways. Enki's shrine stood on an enormous stepped *ziggurat* that was visible many kilometers away.

With the stabilization of sea levels in about 3000 B.C., humanity everywhere entered into a new and infinitely more vulnerable relationship with the environment. The seesaws of post–Ice Age global warming had turned foragers into farmers and villagers into city dwellers, putting them all at the mercy of much shorter cycles of climate change. El Niño and other such short-term climatic episodes assumed a new and menacing importance.

By 3000 B.C., the world was on a trajectory of accelerating population growth that has not eased to this day. The farmers of Abu Hureyra had faced the challenge of population growth and sudden drought by turning to farming while other mammal populations crashed. Farming turned aside a slow-burning population crisis with such success that six thousand years later farming had become the dominant lifeway throughout Europe and Asia and was developing rapidly in the Americas. But the imperative realities of population and carrying capacity rose up again and again, at first in fertile areas like Mesopotamia and Egypt, where river floodplain agriculture and simple irrigation produced several crops a year and much larger food surpluses to support growing numbers of nonfarmers. Technology and, above all, the deployment of human labor forces much larger than a mere household again deflected the crisis of carrying capacity. But the new strategies required not village kin leaders and chiefs, but authoritarian rulers with exceptional, supernatural powers.

We can discern the slow process of village chieftains becoming the great pharaohs in Egypt, where the age-old religious beliefs of the farmer gradually became the ideologies of divine kings, who reigned as living gods and controlled the floods of the Nile. These individuals, with their viziers and lesser officials, became the organizers, the instruments that tamed the natural world with technology and people power. The survival of states and civilizations depended on their abilities, their ideologies, and their capacity to change in the face of sudden climatic change.

Supreme power was a heady elixir for pharaohs and their ilk who presided over the world's preindustrial civilizations. They developed compelling ideologies that made them divine kings, the living embodiment of the gods on earth. With divine kingship came great responsibilities for the well-being of their subjects, who accepted social inequality as the price for the rule of a privileged intermediary with the capricious forces of nature. The illusion worked most of the time, until destruction or starvation made the people think their lords had failed them. Then thousands went hungry, and the ruler paid the price. A cautionary tale from ancient Peru makes the point.

A thousand years ago two great royal dynasties ruled over the river valleys of Peru's north coast, known to us from dimly remembered oral traditions recorded by Spanish writers. Between A.D. 700 and 900, a lord named Naymlap landed at the mouth of the Lambayeque River with a fleet of balsa rafts. He brought with him his wife, a greenstone idol, and a retinue of forty officials, saying that he had been sent from afar to govern the land. Naymlap ruled wisely, but in old age he ordered himself entombed at his capital, commanding his children and loyal followers to spread the word that he had sprouted wings and flown away into eternity.

Twelve lords followed Naymlap in a dynasty that presided over more than a dozen Lambayeque Valley settlements. The last ruler was Fempellec. Legend tells how he offended the gods by moving Naymlap's idol from its ancient resting place. They retaliated with heavy rains, which fell for thirty days and thirty nights. Enormous floods swept away the people's fields. In savage anger, his subjects rebeled, bound Fempellec hand and foot, and cast him into the raging Pacific. Famine and pestilence descended on the land, and the armies

of the Chimu dynasty of nearby Chan Chan conquered their weakened rival.

The Lambayeque Valley was home to at least five cities after
Naymlap's time, among them the ancestral center at Chotuna, which
may have bound together a loose confederation of ethnic centers
linked by alliances and kinship ties. The largest is Batan Grande, a
once-prosperous city whose adobe pyramids tower over tens of thousands of looted graves, many of them once rich in gold ornaments
and full of human sacrifices. A strong El Niño struck in about
A.D. 1100. Deep floodwaters cascaded through the great city, carrying everything before them. Instead of rebuilding, the inhabitants
heaped up piles of wood and brush and burned the pyramids and urban complex to the ground. Perhaps this conflagration reflects the
legend of Fempellec, who angered the forces of nature and brought
destruction to his kingdom.

As the next four chapters show, those who claim they control the
cosmos and the future of civilization survive only as long as they are
able to command the loyalty of their subjects. The lesson is simple:
The ultimate equation of history balances the needs of the population
and the carrying capacity of the land. When carrying capacity is exceeded and technology or social engineering cannot restore the balance, all humanity can do is disperse—if there is the space to do so.

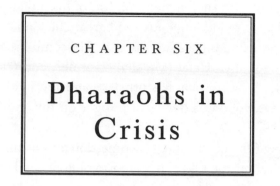

CHAPTER SIX

Pharaohs in Crisis

*If he [the Nile] is sluggish, the nostrils are stopped up, and every-
body is poor. If there be a cutting down in the food-offerings of the
gods, then a million men perish among mortals, covetousness is
practiced, the entire land is in a fury, and great and small are on
the execution-block. [But] when he rises, the land is in jubilation.*
—*Middle Kingdom Hymn to the Nile*

Between 2180 and 2160 B.C., unprecedented droughts in southern
Egypt brought starvation and political disorder to the land. The
famines were so memorable that high officials, such as Ankhtifi, recalled them on their sepulchers:

> All of Upper Egypt was dying of hunger, to such a degree that every-
> one had come to eating his children, but I managed that no one died
> of hunger in this nome [province]. I made a loan of grain to Upper
> Egypt. . . . I kept alive the House of Elephantine during these years, af-
> ter the towns of Hefat and Hormer had been satisfied. . . . The entire
> country had become like a starved (?) grasshopper, with people going
> to the north and south [in search of grain], but I never permitted it to
> happen that anyone had to embark from this to another nome.[1]

Ankhtifi of Hierakonpolis and Edfu (two of the southernmost nomes of Upper Egypt) had such a high opinion of his military abilities that he called himself grandiloquently the "great chieftain." He became a nomarch as low floods beset his districts. His tomb inscriptions boast that "I fed/kept alive Hefat (Mo'alla), Hormer, and (?) . . . at a time when the sky was [in] clouds/storm . . . and the land was in wind, [and when everyone was dying] of hunger on this sandbank of Hell of Apophis."[2]

Memories of the great droughts come down to us through a much-faded historical mirror. Old Kingdom Egypt flourished more than four thousand years ago. Virtually the only documents that survive are tomb inscriptions, chiseled in stone according to prescribed formula. Self-serving and boastful, they glorify their owners to eternity. All we can do is read between the lines and hope to glean some historical truth from the funerary rhetoric. The nomarch Ankhtifi's inscription has a powerful ring of truth: cannibalism, rampant hunger, sandbanks in midriver. His account speaks of hard-won experience and desperate struggle. Some of the behavior Ankhtifi describes is eerily familiar from India's great hungers—the aimless wandering of starving people in search of food, the fighting over precious water supplies.

The memories endured for generations. A vivid account of Egypt in crisis comes from the writings of a sage named Ipuwer, who seems to have suffered through some of the misery and set down his memories later. He painted a graphic picture of Egypt gripped by repeated famines, where "the plunderer is everywhere and the servant takes what he finds." When the Nile did flood, many farmers lacked the confidence to sow crops in a time of chronic uncertainty. Birth rates declined (just as they did during nineteenth-century Indian famines), the dead were thrown into the Nile, and plague swept through the kingdom. Upper Egypt became "an empty/dry waste" as sand dunes blew onto the floodplain from the encroaching desert. "Nay, but men feed on herbs and drink water; neither fruit nor herbage can be found any longer." People attacked and looted the state granaries: "The storehouse is empty and its keeper lies stretched on the ground . . . the grain of Egypt is common property." Tomb robbing was rampant, and "the land has been deprived of kingship by a few lawless men." In a memorable passage, Ipuwer blames the pharaoh of the day: "Authority, Knowledge, and Truth are with you, yet confusion is what you set through the land, also the noise of tumult."

In 1983, while visiting Egypt's major archaeological sites, I had occasion to ride a lateen-sailed cargo boat up the Nile from Thebes to Aswan. We sailed majestically up the great river before the winter north wind. With sudden hindsight, I watched the timeless Egypt of the pharaohs unfold along the riverbanks. A farmer turned the soil with a plow like that depicted in Old Kingdom tomb paintings. White egrets trod the furrows behind him; a man slung from the trunk of a palm tree lowered ripe dates to his wife far below. My mind traveled back to the Egypt of four thousand years ago, when the remote ancestors of these same Nile villagers planted and harvested in a comfortable, seemingly unchanging world.

Ancient Egypt was always a rural state, an intricate patchwork of hamlets, villages, and small towns. The pharaoh and his nobles owned large country estates, as did the major temples, but the breadbasket of Egypt's dazzling civilization rested firmly in the hands of provincial leaders called nomarchs, who presided over hundreds of local communities from Aswan to the delta. Each summer, as the Nile swelled over its banks, small groups of villagers stood watch over their levees, clearing canals and ditches, frantically shoring up collapsing dikes to steer the water to the right drainage basin or storage reservoir. Then, as the inundation receded, men and women plowed and planted the emerging fields. Rich silt left by the river nourished the growing crops. Everyone, pharaoh, noble, and commoner, depended on the bounty of the villagers' fields. The entire state ran on the rations of bread and grain parceled out by an army of scribes and petty officials. When the flood failed and crop yields declined, farmers and nomarchs looked after themselves first and the state second, a lesson that later pharaohs never forgot.

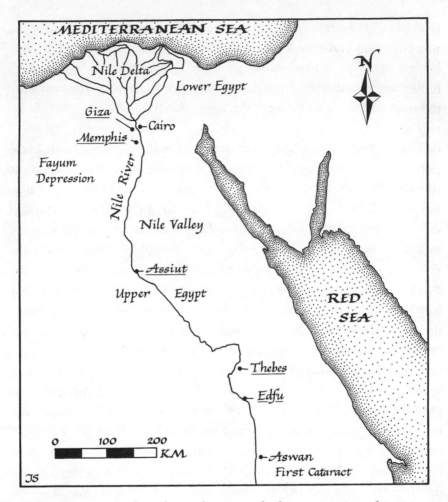

FIGURE 6.1   Archaeological sites and places mentioned in Chapter 6.

The Old Kingdom pharaohs presided over a fertile land carved from the harshest of deserts. Their linear kingdom lay in a 1,300-kilometer-long belt of green land, narrowing and widening with the valley as it followed the meanderings of the Nile. The green arrow of Ancient Egypt pointed northward from the First Cataract at Aswan, as the river waters moved slowly toward the Mediterranean Sea. The Nile was the main artery and channel of communication, with canals and side streams flowing like veins into the densely cultivated lands on either bank. Below where the modern city of Cairo now lies, the river expanded into the luxuriant, fan-shaped delta. For five thousand years virtually no rain has fallen in this narrow land. Egypt depends entirely on rainfall from thousands of miles to the south.

The Egyptians called their homeland *kmt*, "the black land." They considered the great river the source of life, a divine stream that was part of the cosmic order. Each year, the Nile rose and flooded the rich floodplain as runoff from summer tropical rains deep in Africa flowed into the heart of Egypt. This was *Akhet*, the season of inundation, when the river swelled above its banks, turning the valley into a vast, shallow lake. Towns and villages became small islands as the river waters dropped fresh silt on the fields, then receded slowly. Wrote the British irrigation expert William Willcocks, who worked in Egypt in the 1890s: "The Nile looms very large before every Egyptian, and with reason." He recalls his deep sense of awe on seeing "the river rising out of the deserts and flowing between deserts, rising and swelling until it overflowed its valley, while overhead we had a cloudless sky under a burning sun."[3] Then came *Peret*, the season of planting, when barley and wheat ripened slowly in the winter sun without the need for watering. After the harvest in March or April, *Shemu*, the season of drought, descended on the valley. The early summer sun hardened and cracked the ground, aerating the soil and preventing the accumulation of harmful salts. *Shemu* ended with the coming of the new inundation, completing the cycle of Nile farming life.

The Nile floodplain accumulated over many thousands of years as a result of suspended sediment accumulating with the annual flood.[4] The valley itself is of a classic, convex type common to many African rivers: With the lowest levels at the *edges* of the valley, the floodwaters flowed far outward on either side of the channel. Active and abandoned levees of the Nile and its branching channels divided the alluvial basins into natural flood basins. These levees rise between one and three meters above the lowest alluvium, creating shallow basins up to one hundred square kilometers in extent. The farmers raised and reinforced the natural levees, which served as longitudinal dikes for containing the river. Low, sinuous rises, occasional transverse dikes, and humanly dug canals criss-crossed the flat lands.

Under ideal conditions, the Nile would rise to bank-full stage in Upper Egypt by mid-August and then spread through major and minor overflow channels or by breaches across low levees and spill over into successive flood basins. As the surge carried downstream, it would flood the last basins at the northernmost end of Egypt four to six weeks later. At the height of a normal flood, all but the crests of the levees would be briefly flooded, with about one and a half meters of water in the basins. After several weeks, or even months, the flood basins would emerge from the receding waters. Planting would begin in Upper Egypt in September, six weeks later in the delta.

The average floodplain relief was only about two meters, so in good flood years the water would spill over the higher levees. In exceptional years, the inundation would carry all before it in scouring destruction. The tomb robber–archaeologist Giovanni Belzoni witnessed an exceptional flood in 1818, when rapid-running water invaded village after village, leveling houses and the earth-and-reed fences that protected them. He wrote: "Fortunate was he who could reach a high ground. Some crossed the water on pieces of wood, some on buffaloes or cows, and others with reeds tied up in large bundles."[5] The villagers frantically transported their precious grain to higher terrain so they would have something to plant as the flood receded.

A weak inundation, however, would cover only a small part of the plain. Sometimes the flood stage was so brief that the water began to recede almost at once, with catastrophic effects. In A.D. 967, 600,000 people died of starvation, one-fourth of the entire population of Islamic Egypt. During another famine in A.D. 1220–1201, between one hundred and five hundred people a day perished in Cairo alone. A rise that was two meters below average could leave up to three-quarters of some Upper Egyptian provinces totally unirrigated. In Ancient Egypt, low flood years could be disastrous for the king, who depended on grain from farmers to support the court and public works such as pyramid building. Even if the ruler had accumulated enough grain to distribute in hungry years, no amount of grain or storage capacity could enable a ruler to endure several weak inundations without authorizing massive new irrigation works or new farming strategies. A central government is always in a weak position when food is scarce and tribute is late, especially when powerful local officials decide to keep such tax resources as there are for themselves.

One cannot entirely blame generations of kings for their short-sightedness. The Egyptian population was still relatively small, and there was plenty of fertile land to go around. A normal flood allowed a good crop season over some two-thirds of the floodplain. Long before the first pharaoh, the Egyptians had resorted to limited irrigation to feed a growing farming population, a practice that expanded the area of natural crop land that could be cultivated. Dams and reservoirs also allowed the farmer to retain water in natural flood basins after unusually short flood crests. By deploying a relatively small number of people, a village could dredge and deepen natural overflow channels, dig short ditches to breach low points of natural levees, block off gathering streams with earthen dams, and use buckets to raise water from natural ponds or channels into neighboring fields.

The Egyptians knew the Nile was the source of life. Their kings fostered the belief that they controlled the mysterious inundation, the very fountain of human existence. The reality was quite the reverse. Old Kingdom Egypt, with its scant rainfall but ambitious splendor, ultimately depended on the Indian Ocean monsoons and the whims of the Southern Oscillation. In a state where village farming and food supplies were decentralized and self-sustaining, those closest to the land held the key to the pharaoh's survival.

Scientists have long known that droughts in Egypt often coincide with dry conditions in India. In 1908 the *Imperial Gazetteer of India* reported that "it is now fully established that years of drought in western or northwestern India are almost invariably years of low Nile flood. The relation is further confirmed by the fact that years of heavier rain than usual in western India are also years of high Nile flood."[6] While serving in India, Sir Gilbert Walker not only identified the Southern Oscillation but also studied the effects of its pressure seesaw on the eastern side of the Indian Ocean. Walker observed the Nile drought correlation over many years, although he and later researchers were careful to point out that this was little more than a very general observation, made with an awareness that local conditions vary a great deal.

The Nile begins as four rivers, two rising in East Africa, and two in Ethiopia. Far to the south, in Uganda, the Victoria Nile flows from the lake of that name into Lake Albert to the northwest. Then the White Nile flows northward from Albert to Khartoum, where it joins the Blue Nile, which rises in the Ethiopian highlands. The combined Niles flow majestically through the desert, swollen even more by the Atbara River, which also originates in Ethiopia's mountains. Every year summer rainfall from tropical Africa floods down the Nile, bringing water to Egypt far downstream. Since most of the annual floodwaters originate in the Ethiopian highlands, the monsoon rains in northeast Africa were the water pump that kept Ancient Egypt running.

Gilbert Walker was correct in linking fluctuations in the Southern Oscillation with the intensity of Nile inundations. A complicated interplay of high and low pressure affects weather conditions in the Ethiopian mountains. The arrival of most rainfall in the highlands between June and September accounts for the timing of the Egyptian inundation. In most summers, a persistent low-pressure system over India and the Arabian Sea brings strong southwesterlies to the Indian Ocean region. The Intertropical Convergence Zone (ITCZ) lies just north of Eritrea, so abundant rains fall in the Ethiopian highlands. The Blue Nile and Atbara Rivers carry the resulting floodwaters downstream. These conditions prevail when the Southern Oscillation Index is high, resulting in low pressure over the Indian Ocean. When the index is low, the convergence zone stays farther south and the large Indian Ocean low-pressure system moves eastward or develops only weakly. The monsoon winds falter or even fail altogether. Drought affects the highlands. Thousands of kilometers to the north, Egypt experiences a low flood.

How long has this seesaw affected Egypt? If the evidence from the Peruvian coast is correct, El Niño circulations have persisted since at least the beginning of Ancient Egyptian civilization in 3000 B.C. Unfortunately, until someone locates tree-ring sequences of sufficient precision to give annual rainfall fluctuations for the Nile Valley, we will never be able to identify Old Kingdom drought cycles other than from contemporary accounts of somewhat dubious reliability. Current year-by-year records go back just over fifteen hundred years. The oceanographer William Quinn has painstakingly reconstructed the fluctuations of Nile floods from Islamic Egypt in A.D. 641 to modern times, using a variety of historical sources. Continuous, scientific Nile flood records date from 1824 and show marked fluctuations, including a large drought in 1899–1900, which coincided with a strong El Niño event (and a catastrophic Indian monsoon failure) in that year.

Quinn's intricate historical work provides some interesting data on the varying frequencies of poor inundations. His figures reveal a total of 178 weak Nile floods, of varying degrees of deficiency, over 901 years, with a weak inundation an average of every five years. The frequency of low years has varied considerably over the centuries. Between A.D. 622 and 999, there were 102 years of poor floods, which occurred just under 28 percent of the time. This rate contrasts with that of 1000 to 1290, when the world's climate was slightly warmer and wetter than today and Nile floods were below average only 8 percent of the time. Thirty-five percent of the inundations between 1694 and 1899 were lower than usual, a period that coincided with the Little Ice Age, when there were several extended periods of paltry inundations (see Chapter 10). Quinn believes that such extended periods of low discharge were characteristic of cooler periods. He points out that the Old Testament oscillation of seven years of plenty followed by seven of famine, predicted by Joseph in Genesis 41, was nothing unusual.

Quinn found that river discharge rates varied dramatically. The average Nile discharge at the First Cataract at Aswan in 1879 was 129 billion cubic meters per day. In 1913 the discharge fell to just 44 billion cubic meters a day. The lowest in the late twentieth century was in 1972–1973 and 1982–1984, when it was 62 billion cubic meters. Modern Egypt will probably need a minimum of 63 billion cubic meters a day in the year 2000, so the danger of drought still exists. Even after the building of the Aswan High Dam in the 1960s, several years of high flood gave way to panic in 1987, when water levels in Lake Nasser behind the dam wall fell so low that Egyptians feared a loss of drinking water and electric power.

The fragmentary record of Old Kingdom flood levels, from ancient lake levels in the Fayum Depression west of the Nile and from pharaonic sources, tells us that Nile floods were exceptionally high just as Egypt became a unified state in about 3000 B.C. There was a rapid decline during the First and Second Dynasties, at which point court officials began recording flood levels with Nilometers and other devices. (The Egyptians measured flood heights with markings on cliffs or with carved columns and developed the art of flood prediction to a high pitch.) As the Old Kingdom reached the height of its glory, Nile discharges continued to decline. A dry lake named Birket Qarun in the Fayum provides evidence for extremely low flood levels after 2180 B.C. If the Egyptians of the day are to be believed, the subsequent drought cycle was very severe indeed.

Four centuries before "the great hunger," in about 2550 B.C., the Old Kingdom ruler Khufu ordered the building of his pyramid sepulcher and mortuary temple on a rocky plateau at Giza on the west bank of the Nile. His Great Pyramid is an astounding monument: 146.6 meters high, with a base length of 230.3 meters. Its four sides slope at a precise angle of 51 degrees 51 feet 40 inches. Khufu's architects oriented the pyramid with simple astronomic observation only 3 feet 6 inches off true north. More than 2.3 million stone blocks form his sepulcher, each weighing about 2.5 tons. Originally, white limestone casing stones covered the entire pyramid. This was just a beginning. Khufu's entire mortuary complex comprised three smaller pyramids for his queens, a causeway, two temples, and a satellite pyramid, as well as high officials' tombs.

No one knows exactly how long Khufu reigned, but if he ruled for thirty years, his builders would have had to set in place no less than 230 cubic meters of stone a day. The workers would have had to place an average-sized block in position every two or three minutes during a ten-hour working day. During the quiet inundation season, thousands of workers toiled at pyramid building, hauling huge boulders up earthen ramps on wooden sledges as white-clad scribes kept careful record of the stones moved and the rations issued. By any standards, the building of Khufu's pyramid complex, as well as those of his successors Khafra and Menkaura, was a staggering achievement and a masterpiece of careful administration and the feeding of thousands of skilled and unskilled workers—all for the benefit of a single man, the living embodiment of the Sun God Ra. The pyramids of Giza were the supreme achievement of Old Kingdom Egypt, a civilization founded on the assumption that the pharaoh was the fountain of a prosperous world.

Everything depended on the power and spiritual authority of the pharaoh. According to Egyptian belief, the stars were divine beings, and the ruler was destined to take his place among them. "The king goes to his double. . . . A ladder is set up for him that he may ascend on it," says a spell in a Royal Pyramid Text.[7] Thus, the Old Kingdom pharaohs lavished enormous resources on the building of their pyramids. These sepulchers were symbolic sun rays bursting through the clouds to shine on earth. Many years ago I happened to visit Giza on a cloudy day when the sun glanced only fleetingly through the gray. For just a brief moment in late afternoon, brilliant sun rays burst through the clouds and shone on Khufu's pyramid. The ancient imagery came to life as the bright avenue of sunlight became a powerful ladder to heaven.

Egyptian rulers were living gods, the very essence of the divine order of a prosperous world nourished by a bountiful river. The pharaohs *were* human existence, the embodiment of a unified Upper and Lower Egypt. An Egyptian king at the height of his powers was a blend of force and intelligence, nurture and fear, sustenance and punishment. Centuries after the events described in this chapter, a book of instruction for children written in the fourteenth century B.C. described the New Kingdom pharaoh Amenhotep III as "Perception, which is in your hearts, and his eyes search out everybody. He is Re by whose beams one sees. He illuminates the Two Lands [Upper and Lower Egypt] more than the sun. He makes the Two Lands more verdant than does a high Nile. For he has filled the Two Lands with strength and life."[8]

The Old Kingdom pharaohs lived according to *ma'at,* the spirit of rightness, moderation, and balance, the very essence of Egyptian life. To conform to *ma'at* was to achieve immortality and a prosperous existence in the afterlife. *Ma'at* was cosmic order, justice, and prosperity—civilization itself. The ruler, as the personification of *ma'at,* was thought to exercise a magical control over human existence, and over the life-giving floods of the Nile River. The king could not fail. He was the symbol of equilibrium between the forces of order and chaos, of the stability that sustained Old Kingdom Egypt for centuries.[9]

Ancient Egypt was born when a legendary pharaoh named Menes unified Upper and Lower Egypt into a single kingdom some five thousand years ago. Four centuries of political consolidation followed, until the Old Kingdom flowered brilliantly for more than four hundred years. This was the Egypt of the pyramids, of despotic pharaohs like Khufu who ruled as divine kings. The Old Kingdom collapsed after 2160 B.C. as a generation of hunger and political chaos descended on southern Egypt from Assiut to Elephantine. But in the end, Egyptian civilization survived, strengthened profoundly by the years of suffering, to reach new heights in the Middle and New Kingdoms.

The crisis came at a bad moment. Nearly a century earlier, Pharaoh Pepi II had ascended to the throne in 2278 B.C. at the age of six and ruled for no less than ninety-four years, the longest reign in Egyptian history.[10] Pepi ruled long and wisely at a time when life expectancy was approximately twenty-five to thirty-five, so he brought vast experience to the job. His Old Kingdom Egypt was powerful, wealthy, and probably somewhat complacent. Pepi's predecessors had erected the Pyramids of Giza and forged a mighty kingdom in a desert oasis. He inherited wealth beyond imagination. His court at Memphis in northern Egypt enjoyed a long-established royal monopoly on trade in ivory and tropical products with Nubia upstream of Aswan. The pharaoh also controlled the commerce in timber and other raw materials with the city of Byblos, in what is now Lebanon on the eastern Mediterranean coast.

But Pepi ruled in turbulent times. Thirty years into his reign, a Mesopotamian king, perhaps the famous Sargon of Babylon, sacked Byblos and destroyed a major source of Egyptian royal wealth at one stroke. The setback came at a bad time: The pharaoh was spending large sums on foreign expeditions and worrying about his ambitious provincial governors. These nomarchs were responsible for the taxes, tribute, and military levies demanded by the pharaoh, but their loyalties could be suspect.

Pepi maintained his nomarchs' loyalty by appeasement. As he grew older and less effectual, he bestowed much wealth on his local governors, who now aped the pharaoh and built themselves huge tombs in their homelands. As long as the pharaoh was a strong and decisive ruler, who governed by virtue of his spiritual and political authority, the nomarchs trimmed their sails to the political winds and kept tribute flowing. In the early years of his reign, Pepi II and his officials appear to have had close control of the kingdom. But as the king grew older, he became increasingly detached from the business of government. The succession may have been in dispute, since the pharaoh outlived most of his sons. The nomarchs, having gained in both wealth and status, became bolder, more powerful, and less respectful of Memphis.

Pepi II died in 2184 B.C. He was buried in a magnificent pyramid. Some Egyptologists, however, believe his senior officials went to the afterlife in impoverished sepulchers, as if economic times were hard. His son and successor Merenre II did not last long. Nor did a mysterious Queen Nitocris, "braver than all the men of her time, the most beautiful of all the women, fair with red cheeks." She is said to have committed suicide, if she existed at all. Old Kingdom Egypt finally came to an end in 2181 B.C. amid economic and political chaos. The Greek historian Manetho writes of seventy rulers who reigned for seventy days, more a reflection of chronic instability than historical reality.

Pepi's descendants continued to rule from Memphis, but their domains hardly extended beyond the boundaries of the city. For nearly a thousand years, Egypt had gone from strength to strength, a stable society of great artistic, scientific, and technological achievements. *Ma'at* had seemed unchanging, the very essence of political stability along the Nile. Now the ruler, the personification of *ma'at,* had failed his people. With little warning, the central government collapsed and the fragile unity of Upper and Lower Egypt fell apart. The pharaoh's powers wilted in the face of political unrest, social change, and intensifying famine. Only the most powerful and competent nomarchs prospered, for they controlled local grain supplies. The nomarch Ankhtifi was in no doubt of his power. He boasted on his sepulcher: "I am the beginning and the end of humankind, for my equal has not and will not come into being." Ankhtifi and his colleagues ruled like kings.

We know of their deeds from their tomb inscriptions. Nomarch Khety of Assiut grew up at the royal court. His tomb inscriptions boast of how he learned swimming with the pharaoh's children in the carefree days of his youth. Young Khety grew to manhood in prosperous times when brimming Nile floods irrigated the fertile lands of his hometown. Eventually he became nomarch of Assiut, a tax collector, an administrator, and nominally the king's representative in a province far upstream from the royal court at Memphis. Khety's troubles soon multiplied. A series of catastrophically low floods brought hunger to his people. Fortunately, the nomarch was a proactive administrator. His scribes monitored the flood levels with special pillars. They could predict crop yields with remarkable accuracy on the basis of years of previous experience.

Flood after flood peaked quickly, at much lower levels than usual. Thousands of acres of farmland received no water at all. As the Nile fell to record low levels, the people in desperation planted crops on sandbanks. Khety's sepulcher tells us how he, like Ankhtifi, took drastic measures to protect his people:

> I nourished my town, I acted as [my own] accountant in regard to food (?) and as giver of water in the middle of the day. . . . I made a dam for this town, when Upper Egypt was a desert (?), when no water could be seen. I made [agricultural] highlands out of swamp and caused their inundation to flow over old ruined sites . . . all people who were in thirst drank. . . . I was rich in grain when the land was as a sandbank, and nourished my town by measuring grain.

Another official named Merer, "overseer of the slaughterers of the House of Khuu [probably Edfu]," records that he took care of his family and their irrigation works during the famine. "I shut off all their fields and their mounds in town and in the country, I did not allow their water to inundate for someone else. . . . I caused Upper Egyptian barley to be given to the town and I transported for it a great number of times."

The nomarchs knew their districts well and had strong roots in the countryside. They had learned the hard way that only decisive leadership could alleviate hunger. The villagers did not have enough food to feed themselves, let alone the hundreds of nonfarmers who lived by temples and in towns. In times of severe drought, population densities were too high for what the land could produce, and there were few grain surpluses. Competent nomarchs knew that only draconian measures would succeed. They closed the boundaries of their provinces to outsiders to prevent aimless wandering in search of food—a common reaction to mass hunger. They erected temporary dams at the edges of alluvial flats to retain as much floodwater on the fields as possible. Grain was rationed carefully and distributed to the worst-hit areas. Above all, the nomarchs maintained tight control over their domains, did all they could to prevent panic, and kept a careful eye on their ambitious rivals upstream and downstream. They paid nominal homage to the pharaoh, who, theoretically, was responsible for the Nile floods. In practice, his underlings were the real rulers of Egypt, because they could take short-term measures to feed the hungry and stimulate village agriculture.

A revolving door of kings presided over Memphis, their credibility as divine monarchs fatally undermined by their inability to ensure bountiful floods. Egyptian kingship flourished on the assumption that the pharaoh was the Son of Ra, the Sun God, protected by Horus, the celestial falcon deity. Pharaohs were flood-makers, but when the floods failed to materialize and the right order was in question, the king was held responsible. It may be no coincidence that ruler after ruler arose in Memphis as the "true" Sun God while the nomarchs jockeyed for power among themselves. No pharaoh was strong enough to make the floods come.

The fragile unity of the country splintered. The city of Herakleopolis (near modern Beni Suef) won control of the middle reaches of Egypt between 2160 and 2040 B.C. Herakleopolitan rulers may have actually controlled the entire country for a while, but their reigns were short and their dynasties unstable. By 2040 B.C., a bitter rivalry with the ruling family of Thebes led southern Egypt to split off into a rival kingdom. Constant border clashes marked the frontier as the kings of Thebes tried to conquer their northern rival, helped by a return of more bountiful inundations. After 2134 B.C., three Theban kings, all named Intef, tried repeatedly to overthrow Herakleopolis. Each pushed the Theban boundary farther downstream beyond Abydos, setting the stage for the hard-fought military campaigns of Mentuhotep I, who came to the Theban throne in 2060 B.C. and reigned for half a century. The early years of his reign saw much bitter fighting, culminating in a rebellion by Abydos in 2046. Mentuhotep put the rebellion down with decisive severity, reunited Egypt under his rule, and became known as the "Uniter of the Two Lands."

Thus began the Middle Kingdom, two and a half centuries of prosperity and abundance. There were periods of low floods, but none approached those of the great hunger. Mentuhotep I and his two Eleventh Dynasty successors spent their reigns consolidating a weakened kingdom and rebuilding its agricultural economy. The Twelfth Dynasty began when Amenemhet I ascended the throne in 1991 B.C. Seven kings and a queen presided over an increasingly prosperous Egypt and maintained a strong, centralized government during centuries of exceptionally high floods. Religious dogma no longer made the pharaoh infallible. He was charged by the Sun God Re to establish and enforce *ma'at* on earth, a philosophy that allowed for a king's failure. Middle Kingdom pharaohs saw themselves as shepherds of the people, more concerned with the common welfare than their Old Kingdom predecessors. They were also efficient administrators who imposed a firm bureaucracy on every aspect of Egyptian life.

Pharaoh Senusret III (1878–1841 B.C.) was a great warrior and a man of imposing height. (Manetho tells us he was "4 cubits 3 palms, 2 fingers breadth [over 2 meters] tall."[11]) He used the power and prestige of his office, bolstered by bountiful floods, to break the power of the nomarchs and replace them with administrators under direct royal control. The Middle Kingdom reached the height of its prosperity under his successor Amenemhet III (1929–1895 B.C.), under whose reign a series of exceptionally high floods swept through the kingdom. The pharaoh embarked on extensive irrigation and land reclamation works in the Fayum Depression, west of the Nile. He is said to have cleared accumulated silt from the Hawara Channel, which once connected the depression to the Nile. The Fayum served as a combined floodscape and reservoir where lakes filled during high inundation years, protecting Lower Egypt from even worse flood destruction. As the river fell, excess water drained through the same channel.

The Middle Kingdom pharaohs were different from their serenely confident predecessors who built imposing pyramids and lived remote from their people. Though still buried in great splendor, they never forgot the great drought of 2180 B.C. and knew that their power, as guardians of the natural order of things, depended on the erratic floods of the Nile. Middle Kingdom artists crafted telling portrait statues of Senusret III and Amenemhet III that depict them, perhaps deliberately, not as eternally youthful monarchs but as careworn, middle-aged rulers whose brooding seriousness reflects a profound awareness of their responsibilities. The lessons of history weighed heavily upon them, and their royal prerogatives were much weaker than those of their Old Kingdom predecessors.

The great droughts were not the last to affect the ancient Egyptians. Sometime after 1768 B.C., the period of unusually high floods gave way to a regime of irregular inundations that corresponded to near-modern conditions. Between 1768 and 1745, surviving flood-level records hint at famine and several years of low river levels. Boasted an official named Bebi on his tomb at El Kab: "Whenever a famine came during various/numerous years, I gave grain to my city during each famine."[12] The litany is familiar, but the years of hunger were not continuous. A period of political confusion and uncertainty coincided with the declining floods, which saw another procession of pharaohs prompted in part by uncertainties over the royal succession and by the need for the country to adjust to lower flood levels after years of plentiful water. No less than eighteen kings presided over Egypt between 1768 and 1740 B.C., and even more during the 1600s, while astute viziers served through several reigns and ensured some continuity of government.

For all the political uncertainty, Egypt did not collapse completely this time, for each part of the country depended on the other for economic survival. Although Asian Hyksos princes now ruled over the delta to the north, the people of Thebes still pastured cattle there and imported emmer from the fields of their rivals. Technological innovations helped expand crop yields for a rising population; for instance, the *shaduf,* a simple, lever-operated water-lifting device, allowed farmers to water their fields outside the flood season. And when the Hyksos were expelled and Ahmose I of Thebes assumed the throne in 1570 B.C., Egypt entered on its most glorious imperial centuries.

Egypt survived because the people believed their king had defeated falsehood. They ascribed plenty or famine to their pharaohs, not in the modern sense of accountability, but in the belief that their rulers used their divine and human qualities to influence nature. For their part, the best and most powerful Egyptian kings prospered because they deployed their people to create an organized oasis out of natural bounty, which allowed crop yields far above those of simple village farming. They became shepherds, godlike managers of an agricultural state where only firm administration, centralized governance, and technological ingenuity could ensure the survival of their kingdom and enough food in low flood years. These yields supported a steadily rising urban and rural population, despite occasional years of hunger and crisis.

Many centuries later the rulers of a glittering civilization in a much riskier environment on the other side of the world did not exercise such wisdom.

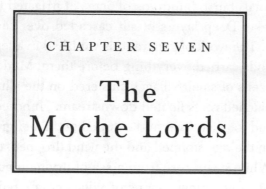

CHAPTER SEVEN

The
Moche Lords

*Rowing up and downe with small reedes on either side, they goe a
league or two into the sea, carrying with them the cordes and
nettes. . . . They cast out their nettes, and do there remaine fishing
the greatest parte of the day and night, untill they have filled up
their measure with which they returne well satisfied.*

—*José de Acosta*, **Historia Natural y Moral de las Indias**, *1588*

Sometime in the mid-sixth century A.D., a strong El Niño brought
torrential rains and catastrophic flooding to the northern coast of
Peru. Thick, black clouds massed offshore, then thickened as they
moved over the densely populated coastal river valleys. Heavy raindrops pattered on the arid ground, cracked and hard from severe
drought. A powerful smell of wet earth permeated the air as the
shower intensified, then stopped abruptly. Ever thicker clouds
massed overhead, mantling the surrounding hilltops. Then the rain
started, carried by a roiling wind from the ocean. Curtains of water
pounded the valley in solid sheets. The rain continued unabated—
mist, steady downpours, intense cloudbursts that flowed down dry
hillsides. Normally placid rivers fueled by mountain runoff burst

their banks and inundated the densely cultivated floodplain. Dikes gave way, canals burst, hundreds of acres of irrigated land became a freshwater lake. Deep layers of silt cascaded over carefully tended field systems. The work of generations vanished in a few hours as the rains and floods carried everything before them. Muddy water overwhelmed dozens of small villages clustered on the alluvium. Houses collapsed, thatched roofs floated downstream. Hundreds drowned as the people fled for their lives and camped on higher ground.

Even when the rain stopped and the wind dropped, the destruction continued. White steam rose from a sea of drying mud where green, irrigated crops once grew. Teams of villagers labored frantically to save untouched fields threatened by rising waters. Hills and valleys were awash. Erosion gullies gashed desert hillsides as millions of tons of sand and river silt swept out to sea. Huge Pacific swells driven by onshore winds pounded the beaches, piling great sand dunes above high-tide levels. The fine sand swirled and blew inland, burying farmland and blocking river valleys. The dunes were mountains of destruction on the move.

Thirty kilometers inland from the raging Pacific, two brightly painted adobe pyramids towered over the inundated Moche Valley. Here the rulers of the glittering Moche state looked down on their highly irrigated and normally well-organized domains. The haughty lords were high above the muddy water, but the rain left its mark on their adobe mountains. Water flowed into tiny cracks in the stucco, turning small imperfections into deep crevices, and crevices into wide rifts, as the clay crumbled. Deep erosion gullies soon cratered the once-smooth sides of the sacred edifice as El Niño mocked the divine powers of the Moche leaders.

The sixth-century El Niño and the droughts of the same century sowed the seeds of destruction for one of ancient America's most spectacular and powerful civilizations. The tragic story of the Moche is a telling indictment of inflexible, despotic leadership.

The Moche civilization flourished along the arid north coast of Peru between A.D. 100 and 800. Thousands of farmers and fisherfolk lived under the rule of a small number of authoritarian warrior-priests. The Moche lords never ruled vast domains, just a strip of some four hundred kilometers of coastal desert between the Lambayeque Valley in the north and the Nepeña Valley in the south. Their subjects dwelt along the Pacific or in dry river valleys that fingered no more than about one hundred kilometers inland through one of the driest environments on earth. The lords themselves lived apart from their subjects, in high palaces set atop massive adobe pyramids that gleamed brightly in the sun. They presided over an orderly, well-organized world, defined by a vivid and still little-known set of religious beliefs. But it was a world plagued by droughts and El Niño.

The warrior-priests lived in a world of their own, far from the daily work in the irrigated fields beneath their magnificent pyramids. As far as we can tell, their lives revolved around warfare, ritual, and diplomacy, in an endless cycle of competition for prestige with their fellow leaders. Each river valley had one or two royal courts, all of them connected by ties of kin and mutual obligation. Judging from royal graves, each warrior-priest wore the same insignia and ceremonial trappings. Moche lords went to war over land and water supplies. Painted Moche pots show vivid scenes of armies fighting with raised clubs and feather-decked shields. Other paintings depict naked prisoners of war paraded before the warrior-priest dressed in his full ceremonial regalia. At a signal, executioners decapitated the captives or strangled them. As the victims choked to death, their penises sometimes became erect in a potent symbol of human fertility. This act of sacrifice in the presence of the warrior-priest validated lordly power over human life. The arrogant Moche lords were the intermediaries between the living and the forces of the spiritual world that could wreak awful havoc on their coastal homeland.

We do not know the names of the first Moche rulers, or any details of their military campaigns or of the diplomatic alliances that linked valley to valley over generations. We do know from their burials that they were people of remarkable wealth and power.

In the late 1980s, the Peruvian archaeologist Walter Alva excavated three unlooted warrior-priest burials from an adobe platform at

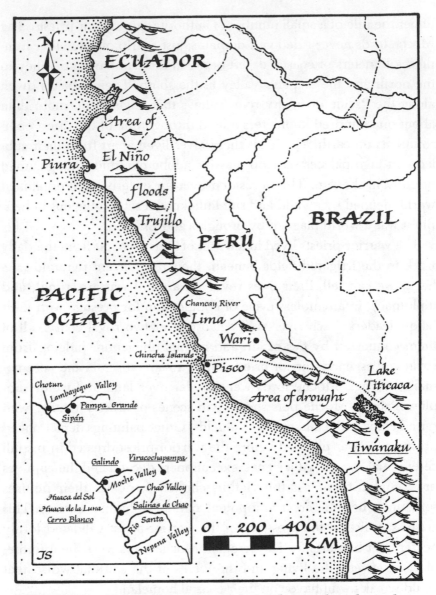

FIGURE 7.1 The Andean area, showing key archaeological sites, places, and cultures mentioned in Chapter 7. Areas of drought and excessive rainfall during ENSO events are also shown.

Sipán in the Lambayeque Valley, center of the Moche world around A.D. 400. Each burial lay inside an adobe brick burial chamber, one placed above the other in a mortuary platform set among the imposing pyramids that rose dramatically out of the heavily cultivated river valley. Each warrior-priest lay in his ceremonial garb. Their regalia never changed from one generation to the next—elaborate gold masks and crescentlike headdresses that once glinted in the bright sun, cotton tunics adorned with dozens of copper gilt plates, bead pectorals and silver or gold ear pendants with hinged figures of warriors—silver symbolizing the moon, gold the sun. Anyone gazing on a warrior-priest shining in golden splendor was left in no doubt of his ability to control the forces of the Moche world.

Royal artisans crafted their masters in clay. The great men gaze haughtily into space with the calm assurance of unquestioned political and spiritual authority. Moche society was a social pyramid, erected on the backs of anonymous villagers who worked in the fields and on irrigation canals, paying a yearly tax in labor for the good of the state. Thousands of villagers labored to build enormous royal pyramids, to construct and maintain canals and irrigation systems to feed their lords and a privileged few. Like Egyptian pharaohs, the Moche lords depended for their power on their ability to provide water, ample crop yields, and food surpluses to tide every village through drought years and floods. Under the powerful religious ideology that linked rulers and the ruled, warrior priests interceded with the spiritual forces that controlled an arid land.

The Moche civilization had two economic pillars. Their brilliant farming expertise harnessed mountain runoff and fertile soils with large irrigation systems capable of producing substantial grain surpluses and acres of cotton for their fine textiles. The floodplains of the Lambayeque, Moche, and other coastal valleys formed green patchwork quilts of closely packed irrigated fields, nourished by long canals. Thousands of hours of cultivation, ditch digging, and maintenance went into the Moche field systems. However, everything depended on careful sharing of mountain runoff, an annual gift in the hands of the spiritual world.

The second pillar was a bountiful Pacific. Coastal upwelling brought swarms of anchovy to feed, and Moche fisherfolk in their reed boats harvested millions of anchovy throughout the year. They dried whole fish in the sun or ground their catch into nutritious fish meal. Thousands of kilograms of protein-rich fish meal traveled from the coast to amplify the carbohydrate diet of highland farmers far in-land. The anchovy, by helping to support a large nonagricultural population, was of vital economic importance to the Moche.

The same anchovy bounty fed enormous numbers of seabirds such as cormorants. In turn, the Moche mined bird guano from offshore islands and spread it as fertilizer on their fields. Rivers, marshes, and freshwater lagoons provided additional food.

In normal times the Moche had an abundant and nutritious diet, with enough food to support high population densities, not only farmers and fisherfolk but the thousands of non-food producers who labored on irrigation works and supervised the building of enormous pyramids, palaces, and temples. Each major center supported skilled artisans who labored in clay, cotton, and metals to produce magnifi-cent artworks and formal regalia for the tiny numbers of people who controlled the destiny of the state.

With such ample food supplies, the Moche could have prospered indefinitely but for three spoilers: drought cycles, earthquakes, and El Niño.

Strong El Niño episodes were so rare that relatively few people witnessed a truly catastrophic incident during their lifetimes. In a so-ciety where life expectancy was twenty-five to thirty-five years, gener-ational memories were short. Great El Niños soon became remote happenings, almost beyond recollection except as near-mythic cata-strophes recalled in oral traditions. The Moche lived in a uniform, usually reliable environment, with only drought as a common haz-ard. They built an elaborate, top-heavy society on this shakiest of foundations. Over many centuries, the coastal valley populations steadily rose, far beyond the natural carrying capacity of the arid environment, until the only way to feed everyone was through massive, highly organized irrigation systems that used every drop of water that cascaded downstream during the spring mountain thaw. The warrior-priests held the reins of power because they could deploy thousands of villagers to build and maintain some of the most intensive irrigation works ever built in the ancient world. But their tight hold on power was transitory, because their inflexible beliefs and insistence on absolutely centralized government left them at the mercy of powerful environmental forces. They had no long-term strategy against a phenomenon as maverick as El Niño.

The Moche's roots lay deep in the ancient cultures that had flourished for millennia along the arid Pacific coast of Peru. For thousands of years, small fishing and foraging groups settled at the mouths of dry coastal river valleys. They lived off the bountiful anchovy shoals that fed close inshore. In years when Pacific waters warmed up and the anchovy left for new feeding grounds, the people simply consumed other foods, such as shellfish and the winter fog–nourished wild plants that grew in dense clumps in river valleys close to the ocean. The ancient coastal people had such a variety of foods to live on that many groups stayed in one place for months on end. Theirs was a flexible, egalitarian society, with a reasonably ample supply of alternative foods to fall back on if the fishery failed or torrential rains came. Although some settlements were larger and more permanent, their total population was but a few thousand, partly because they lacked cotton for nets and lines and hollow gourds to serve as net floats, critical innovations for fishing on a large scale.

Population densities even in the most favorable areas were sufficiently low to allow for a degree of mobility, while stands of wild plant foods had not yet been overexploited. For many centuries, most groups lived in foggy zones close to the Pacific, where such foods were to be found. By 2000 B.C., many had moved. They still gathered plant foods but also relied on domesticated plants such as squashes and beans grown on river valley floors.

Today we take cotton for granted. The Andeans not only domesticated fine-quality, long-stranded cotton but made their mountain and lowland homeland a crucible of agricultural innovation, developing many different strains. Cotton soon became a staple of coastal society, used for fishing nets and lines and as a substitute for textiles made of cactus and grass fiber. Coastal communities discovered that cotton could be grown successfully in low-lying warm environments, and they developed a lucrative trade in fabric with the mountain highlands. This may have been why they opened up large areas of desert valley land to irrigation when many families and neighboring villages joined together to develop small-scale water control works. By 1800 B.C., as Egypt was recovering from political chaos, some Andean north-coast settlements boasted between one thousand and three thousand inhabitants living off a combination of fishing and agriculture. Valley populations rose far above the natural carrying capacity of the desert as the Andeans used irrigation technology and new styles of leadership to feed much larger communities.

Under such circumstances, how does a society cope with a strong El Niño? Given the Andeans' short life expectancy, El Niño strategies must have resided in communal rather than individual memory. As the New Guinea anthropologist Roy Rappaport emphasized as long ago as 1971, religious rituals were vital in the communication and validation of information, especially in areas like the Peruvian coast where conditions could change within a few weeks and rapid, flexible responses were essential. Under these circumstances, individuals with spiritual authority play important roles in supervising food storage. Over the generations, distinctive ideologies developed that validated strong leadership and the complex relations between the living and spiritual worlds. The messages of these ideologies have come down to us on textiles.

The coastal people were expert weavers. The dry climate of the Peruvian coast has preserved cotton cloth well over two thousand years old, much of it dyed in at least 109 hues in several natural color categories. The weavers spelled out a simple ideology in symmetrical, angular motifs that persisted for centuries. Anthropomorphic figures, perhaps shamans in trance, appear with flowing hair or snakes dangling from their waists. Birds of prey, snarling felines, and two-headed snakes hint at close ties with a mythic, supernatural world. When fresh and brightly colored, the cotton textiles vividly depicted the myths, themes, and spirit creatures that inhabited the cosmos. The textiles bore a straightforward message that forged close links between the living and the dead, between communities and their ancestors, who interceded with the beings that controlled the land, food supplies, and forces of nature.

The coastal world changed rapidly as each community became more sedentary, more closely tied to river valley lands. By 1000 B.C., the kin leaders of earlier centuries had parlayed their powers as intermediaries with the ancestors into new, powerful roles. The ablest of them became competitive, authoritarian chieftains, capable of marshaling hundreds of villagers to labor on impressive pyramids and mounds honoring the supernatural forces that controlled the Andean world. During the quiet months of the farming year, teams of kin-group members labored to build such structures of rubble and adobe, some of which would endure for more than three thousand years. The Salinas de Chao pyramid on the north coast stands more than twenty-four meters high, a dwarf by the standards of Egypt's Pyramids of Giza, but still an imposing structure. The foundations of Moche civilization stemmed from these earlier cultural traditions, according to which shaman-rulers commanded by virtue of their control over the supernatural world.

The Moche state, named after the small river valley where it began, coalesced out of smaller valley kingdoms at about the time of Christ. The new supreme rulers took all the reins of power into their hands as they crafted a highly centralized state. Their capital, at Cerro Blanco in the heart of the Moche Valley, was a center of immense wealth and spiritual authority. The city's magnificent buildings sprawled around two enormous adobe *huacas* (pyramids). Nearby lay extensive cemeteries and the many workshops that produced ceremonial pottery and other artifacts for the elite.

Huaca del Sol and Huaca de la Luna formed sacred mountains towering over the city. Here the Moche lords had their palaces and conducted solemn ceremonies, some of them involving human sacrifice. Huaca del Sol rises forty-one meters above the plain, two-thirds the height of the contemporary Pyramid of the Sun at the Mexican city of Teotihuacán. The vast pyramid formed a giant cross, with the front facing north. Four sections created a steplike effect, the lowest supporting a ramp, now vanished, that led to the summit. If Spanish records are to be believed, the third stage was a royal burial place, and the fourth and highest the palace of the supreme ruler.

Huaca de la Luna is a complex of three platforms that were once interconnected and joined by high adobe walls. The walls were decorated with richly colored murals of anthropomorphic and zoomorphic beings, of animated clubs, shields, and other artifacts, themes sometimes repeated on Moche ceremonial pottery. The two vast *huacas* reflect the duality of Moche rule: The human divine ruler resided at one *huaca,* the imperial pantheon at the other.

From the distance of centuries, the Moche state gleams with brilliant artistic achievement and all the trappings of almost monolithic stability. The massive mud-brick pyramids of its powerful lords loom above barren hillsides or in the midst of intensively cultivated farmland. The brilliance was indeed real, but it was a transitory luster, maintained at enormous cost in a climate where water was a priceless commodity. The Moche's foundations, built on mountain runoff, river sand, and ocean upwelling, were always vulnerable.

The theory that El Niños helped cause the demise of Moche civilization comes from a new understanding of the vagaries of Andean climate, obtained from snow accumulation records found deep in mountain glaciers.

The Quelccaya ice cap in the Cordillera Occidental of the southern Peruvian highlands lies in the same zone of seasonal rainfall as the mountains above Moche country. Two ice cores drilled in the summit of the ice cap in 1983 have provided a record of variations in rainfall over fifteen hundred years and, indirectly, an impression of the amount of runoff that would have reached lowland river valleys during cycles of wet and dry years. El Niño episodes have been tied to intense short-term droughts in the southern highlands, as well as on the nearby *altiplano,* the high-altitude plains around Lake Titicaca. The appearance of such drought events in the ice cores may reflect strong El Niño episodes in the remote past. However, it is more productive to look at long-term dry and wet cycles, which were part of the potentially catastrophic fate awaiting Moche civilization.

Each of the two ice cores, 154.8 and 163.6 meters long, yielded clear layering and annual dust layers that reflected the yearly cycle of wet and dry seasons, the latter bringing dust particles from the arid lands to the west to the high Andes. The average year-round temperature at Quelccaya is minus three degrees Celsius—so far below freezing that the annual variations in the ice core accumulation reflect actual precipitation rather than variations in the intensity of summer melting. The research team believes this was the case for all of the fifteen hundred years of the cores, and that the annual rings give a chronology that is accurate to within about twenty years.

What the researchers call the "accumulation record" shows clear indications of long-term rainfall variations. A short drought occurred between A.D. 534 and 540. Then, between A.D. 563 and 594, a three-decade drought cycle settled over the mountains and lowlands, with annual rainfall as much as 30 percent below normal. Abundant rainfall resumed in 602, giving way to another drought between A.D. 636 and 645. The ice core data is still being refined, but the best drought records come from the lower portions of the ice accumulation record, where the late-sixth-century drought cycle qualifies as exceptionally severe.

The thirty-year drought of A.D. 563 to 594 drastically reduced the amount of runoff reaching coastal communities. We can gain some idea of the severity of the effects from modern water consumption figures. Over the forty-nine years from 1937 to 1985, local coastal farmers used an average of 88 percent of the runoff in the Lambayeque River. Even allowing for the intensive farming of water-hungry sugar cane today, the Moche usage figure would still be high. The effect of a 25 or 30 percent reduction in the water supply would be catastrophic, especially for farmers near the coast, well downstream from the mountains.

How did these dramatic rainfall shifts and El Niño events affect Moche civilization? The lords of Sipán ruled over a portion of the Lambayeque Valley around A.D. 400, soon after political power had shifted northward. Moche society apparently prospered until the mid-sixth century's severe drought cycle. At this time the Moche lords lived downstream, as close to the Pacific as to the mountains so that they could control both water and fisheries. Their power depended on their ability to exercise strict supervision over all food supplies, over every load of fish meal, dried seaweed, and cotton that traveled to the distant highlands. Above all, they watched the life-giving rivers with sedulous care.

Centuries later, Spanish chroniclers wrote of a centuries-old tradition honored by coastal Chimu lords, who ruled over a powerful north-coast kingdom conquered by the Inca in the 1560s. The Chimu allowed the farmers living the farthest downstream to water their fields first. Their Moche predecessors probably followed the same practice, which made economic and political sense, partly because the most powerful rulers lived downstream. As the sixth-century drought intensified, this long-established policy faltered. The diminished runoff barely watered the rich farmland far downstream. No water reached marginal fields at the edges of the valleys. Kilometers of laboriously maintained irrigation canals remained dry. Blowing sand cascaded into empty ditches. By the third or fourth year, as the drought lowered the water table far below normal, thousands of acres of farmland received so weak a river flow that unflushed salt accumu-

lated in the soil. Crops withered. For the first few years, the lords retained firm control by carefully husbanding the grain supplies stored
by the state against lean years. Fortunately, the Pacific fisheries still
provided ample fish meal—until El Niño came along with unpredictable irregularity.

We do not know the exact years during the long drought when
strong El Niños struck, but we can be certain that they did. We can
also be sure they hit at a moment when Moche civilization was in crisis, with grain supplies running low, irrigation systems sadly depleted, malnutrition widespread, and confidence in the rulers' divine
powers much diminished. The warmer waters of the Christmas Child
now reduced anchovy harvests in many places, decimating a staple of
both the coastal diet and the highland trade. Torrential rains
swamped the Andes and coastal plain. The arid rivers became raging
torrents, carrying everything before them. Levees and canals overflowed and collapsed. The arduous labors of years vanished in a few
weeks. Hundreds died. The survivors moved to higher ground where
they camped under temporary shacks.

El Niño devastated the valley landscape. Dozens of villages vanished under mud and debris as the farmers' cane and adobe houses
collapsed and their occupants drowned. The floods polluted springs
and streams, overwhelmed sanitation systems, and stripped thousands of acres of fertile soil. As the water receded and the rivers went
down, typhoid and other epidemics swept through the valleys, wip-
ing out entire communities. Infant mortality soared. Cerro Blanco
suffered badly. Floodwaters covered parts of the city with as much as
ten meters of alluvium, stripping away five meters of soil or more in
other places. The river eroded away much of the *huacas,* turning them
into crater-sided hills.

The valley soils were so dry that much of the surface water soon
vanished into the ground, turning the devastated floodplain into
hardened mud. The coastal fisherfolk did not starve, for they could
subsist off the unfamiliar tropical fish that now visited coastal waters.
However, even a full-strength anchovy harvest could not feed the en-

tire Moche farming population. We can be certain that several years of hunger gripped the kingdom, the flood damage aggravated by the ongoing drought. All the rulers could do was to muster the enormous, if weakened, labor force at their disposal. Teams of villagers set to work building entirely new irrigation systems adapted to tortuously changed topography. Some of these canals and field systems can still be traced on the ground today. Meanwhile, the lords sought assistance from distant allies in the highlands to the east, who had developed more drought-resistant forms of maize, with large cobs and more kernels. The new strains appear in archaeological sites dated to soon after the flood.

The Moche lords sat at the pinnacle of a state in crisis, their palaces and pyramids ravaged by raging floodwaters. Hedged in by centuries of religious ideology that rationalized their supreme power, they lived apart from their subjects. Unlike European monarchs, who passed their wealth and power to their descendants, the Moche rulers took their rich possessions to the grave, so their successors had to commission new ornaments and opulent clothing from the royal artisans clustered near their palaces. Continuity came not from wealth but from a lord's ability to command the loyalty of his subjects and thus to control agriculture, trade, and water supplies. The El Niño–caused disaster at Cerro Blanco undermined the confidence of generations.

The capital weathered the flood. Huaca del Sol and Huaca de la Luna ascended on heightened platforms. Hundreds of workers labored to build a new and imposing facade to mask the flood damage. One polychrome mural on the walls of one of the Huaca de la Luna courts now bore the figure of a front-facing anthropomorphic being. The figure carried a vertical staff in each hand, which reminds the archaeologist Michael Moseley of a similar central figure that adorns the famous Gateway of the Sun at Tiwanaku, close to Lake Titicaca in the distant southern highlands. The mural soon vanished under a later wall painting. Moseley believes the brief appearance of the staff figure is a sign of highland influence in the religious order, as if the people had lost confidence in their own lordly ideology.

For a while the capital regained some of its former greatness. Then a new disaster threatened, perhaps triggered by yet another strong El Niño. Huge sand dunes, formed from river sediments washed ashore by ocean currents, now blew inland and threatened precious farmland. The advancing sand hills covered hundreds of acres of farmland on the southern side of the valley, then inundated the capital itself. Between A.D. 550 and 600, Cerro Blanco's warrior-priests abandoned their capital and moved inland to the strategic neck of the valley, where the Moche flowed out onto the coastal plain. Almost simultaneously, as if in a concerted decision, the lords of the Lambayeque Valley also moved upstream.

Like Egyptian pharaohs and their courts, the Moche nobility lived out their lives in a time capsule, far distanced from the grinding lives of farmers and fisherfolk. Noble families intermarried, competed, quarreled, and went to war. In times of disaster, these close links facilitated coordinated decisions that affected tens of thousands of people. Such a moment may have come when the besieged lords of the Moche and Lambayeque Valleys decided to move their capitals far upstream to the river necks. By settling in such locations, the rulers could control water supplies, not only to conserve them but also to defend them against aggressive highland neighbors. Sadly diminished, but with at least a semblance of their former prestige and wealth intact, the warrior-priests presided over a much-reduced kingdom. Their people huddled around them, living as close to the water source as possible, far away from the relentless sand dunes that menaced the lands downstream.

Dunes may always have been a problem for farming villages close to the coast, for their shifting sands would bury acres of fertile land quickly and permanently. The problem intensified during strong El Niños, when dune formation could become overwhelming, especially after earthquakes had deposited masses of hillside debris in river valley floors.

We know this because the space shuttle *Challenger* photographed a young beach ridge at the mouth of the Río Santa region, eighty kilometers south of the Moche River, in 1983. NASA scientists digitized seven earlier aerial and satellite photographs of the same beach and were able to show that the new ridge formed between 1970 and 1975 as a result of two natural disasters. A 7.7 Richter scale earthquake in 1970 killed more than sixty thousand people and generated huge landslides in the adjacent dry mountains. The strong El Niño of 1972–1973 brought torrential rains that carried the earthquake debris downstream and deposited it at the mouth of the Santos River. The flood-swollen stream carried the sediment into the ocean, where strong currents pushed it along the beach. Then wave action formed a new sand ridge on the beach. Soon, marching sand dunes crept inland as strong winds blew off the Pacific, burying hundreds of hectares of farmland. Ten years later the 1982–1983 El Niño brought even more earthquake debris downstream. The dunes grew even faster. We cannot detect ancient earthquakes with any accuracy, but the combination of a major earthquake, drought, and a series of strong El Niños may have been the trigger for widespread famine and political upheaval on several occasions in the past.

The Moche rulers consolidated their settlements at the necks of coastal rivers between A.D. 550 and 600. Galindo became the largest settlement in the Moche Valley. Twenty kilometers inland, the new valley capital covered five square kilometers, placed strategically to control the irrigation systems below the neck as well as to give access to lusher, upper-valley farmland. By earlier Moche Valley standards, Galindo was a modest capital, with much smaller platform mounds and less imposing adobe residences. The nobility lived in elaborate houses close by adobe-walled compounds, but interestingly, no less than one-fifth of the town comprised storage facilities, including stone-lined bins built on terraced hillsides flanking the adobe enclosures. The new compounds for the elite and a much greater concern with control of food supplies hint at a society with strong secular concerns, one in which power rested on the ability to control water and to distribute food in times of hunger.

The lords of the Lambayeque Valley founded a new center at Pampa Grande, 65 kilometers from the Pacific and 165 kilometers north of Galindo. This strategically based royal capital controlled the major intakes of all local regional canals. Much of the valley population crammed into a rapidly growing settlement that covered six square kilometers. The archaeologist Izumi Shimada estimates that between 10,000 and 12,500 people lived in the city, many of them farmers who commuted to the fertile fields immediately below the river valley necks. At Pampa Grande, a large canal drew water from the Chancay River and carried it about two kilometers downstream into the canalized Lambayeque. Another maximum-elevation canal captured runoff from the Chancay four kilometers upstream of Pampa Grande. Everything worked toward central control, both of precious irrigation water and field systems and of food surpluses, and was rationalized with traditional religious iconography, carefully reflected in prestigious art objects and architecture.

The religious and political leaders of Pampa Grande lived at the north end of the capital, where the imposing Huaca Fortaleza rises thirty-eight meters above the valley floor. This massive structure, with its long access ramp and surrounding adobe compound, was built exactly like the earlier *huacas* downstream. Once again a highly centralized administration controlled water supplies, grain storage, trade, and the production of craft goods of every kind.

To move upstream and intensify control of irrigation works and farming were unprecedented shifts that may have tested the rulers' leadership to the full. Nothing in earlier Moche history prepares one for this drastic move, but it was not enough. Fixed in their ideological mind-set, the Moche rulers seem to have dismissed more flexible approaches to farming that allowed for both drought and flood. Obsessed with reliable water supplies, they settled near the intakes for major irrigation canals that linked one valley with another and watered the fertile land nearby. This conservative, long-term strategy had the advantage of providing the maximum amount of water in drought years, while anything extra in a rainy year was a bonus.

However, it left the Moche rulers' subjects even more vulnerable to catastrophic flood, and also to their ambitious highland neighbors.

After A.D. 500, prolonged droughts in the southern highlands far from the Moche homeland coincided with strong El Niño episodes in the north, causing widespread political adjustments as different groups, rulers, and kingdoms competed for farmland and control of water supplies. New states arose in the highlands that adopted an entirely different, and arguably more effective, approach to economic and political control of their extensive domains. Whereas the Moche had always governed local populations through highly centralized organizations, the new highland states, such as the Wari, decentralized their governance; a hierarchy of leaders lived among the widely separated populations they administered. The advantage of this system—islands of closely controlled local communities living in valleys and other strategic locations—was that a highland kingdom had a widespread economic reach, each node being linked to the others by tax and tribute obligations. Good road and caravan communications, a long tradition in highland Andean society, were essential. Force, ideology, and careful use of conspicuous privileges tied together dozens of communities in a flexible political map that brought resources from thousands of square kilometers to the nobility with minimal investment in outlying areas. The Wari flourished in these ever changing conditions, and the tentacles of their kingdom expanded and threatened the weakened Moche domains.

Why did the Wari succeed while the Moche staggered under environmental stresses? The center of the Wari kingdom lay on a hilly plateau twenty-five kilometers north of the modern highland city of Ayacucho. Between twenty thousand and thirty thousand people lived in the sprawling capital of stone buildings. Wari farmers cultivated potatoes and root crops at high elevations where mountain rainfall watered their fields. At the same time, they developed highly ingenious irrigation systems that used water sources high above the city linked to downslope field systems by a long primary canal that followed the contours to much lower canals and ditches. This gravity-fed system required a tremendous investment of labor to build and maintain, but it insured the Wari against droughts by drawing water from much more predictable high-altitude water sources. Their vast reclamation technology enabled them to cultivate an enormous area of the Ayacucho region with its steep hillsides. At the same time, the farmers' experiments with new, high-yielding varieties of maize produced large food surpluses.

The Wari knew a good thing when they had one. They exported their new agricultural technology like modern entrepreneurs, packaging it like a computer system complete with its organizational structure of leaders and officials, religious beliefs, and art styles. Significantly, the Wari iconography included the ancient staff-bearing Gateway God from Tiwanaku in the southern highlands, painted in Moche style at Huaca de la Luna at Cerro Blanco during the political crisis brought on by the sixth century's thirty-year drought.

Before and during the great drought, the Wari's unique agricultural technology gave them a competitive advantage over their neighbors. The rapidly growing state exported its agriculture, ideology, and leadership to other areas of the highlands and encroached on the lowlands. Wari leaders may have conquered some valleys by force, but their highly flexible adaptation to unpredictable weather conditions probably gave them a competitive edge, especially against their struggling Moche neighbors. As the successful agricultural technology spread, populations rose. The Wari established a presence far to the north, presumably to foster trade with the north coast and to ensure regular supplies of sacred *Spondylus* tropical seashells, which had great ritual importance in the highlands. In about A.D. 650, they built an imposing provincial center at Viracochapampa, some 170 kilometers south of Pampa Grande, but never occupied the compound. We do not know whether the Wari ever fought the Moche for control of their precious water supplies. However, Moche graves from the Pampa Grande region contain clay vessels that bear Wari motifs, as if highland religious ideas had spread into the lowlands.

With their conservative water conservation and storage policies, the Moche lords governed with none of the flexibility of their highland neighbors. Everything depended on the green irrigation systems within sight and downstream of their carefully administered river necks. With shattering inevitability, torrential El Niño–borne rains brought the concentrated fury of floodwaters from the entire Lambayeque and Moche watersheds to bear on the carefully engineered feeder canals and irrigation systems.

In the late seventh century, an exceptionally severe El Niño washed away many of the field systems around Galindo and Pampa Grande. Another drought settled on the north coast, intensifying already critical food shortages. Internal unrest, even attacks from hostile Wari, ensued as a weakened leadership wrestled with famine and lost spiritual credibility. Around A.D. 700, the Moche abandoned both settlements. According to Shimada, Pampa Grande came to an abrupt end. Flames engulfed the Huaca Fortaleza, turning the adobes on the summit into fired brick. Whether this destruction was a deliberate act or carried out in the heat of battle, we do not know. Whatever the cause, repeated El Niño flooding and drought had broken the back of a wealthy and powerful state.

Like Egypt's pharaohs, the Moche lords survived at least one disastrous episode that nearly imploded their kingdom. But unlike the Egyptians, they did not learn from their mistakes in an environment that offered nothing like the flexibility and potential of the Nile Valley. Their elaborate irrigation systems created an artificial landscape that supported dense farming populations in the midst of one of the driest deserts on earth, where farming would be impossible without technological ingenuity. The farmers were well aware of the hazards of droughts and El Niños, but technology and irrigation could not guarantee the survival of a highly centralized society driven by ideology as much as pragmatic concerns. There were limits to the climatic shifts Moche civilization could absorb. Ultimately, the warrior-priests ran out of options and their civilization collapsed.

CHAPTER EIGHT

The
Classic Maya
Collapse

*Such was the scattering of the work, the human design. The people
were ground down, overthrown. The mouths and faces of them
were destroyed and crushed.*
–Popol Vuh, The Quiche Maya Book of Counsel

Lake bed deposits in Mexico's Quitana Roo tell the same story as the
Quelccaya ice cap in the Andes. A severe drought cycle settled over
Peru and Central America in the mid to late sixth century A.D.
A combination of drought and El Niños nearly destroyed the Moche.
Years of drought in the tropical southern lowlands of Guatemala and
Mexico caused economic and social disruption at a time of rapid
population growth. Like the Moche, the Maya survived, but the envi-
ronmental writing was on the wall. Three centuries later their civiliza-
tion lay in ruins.

The ancient Maya were among the most flamboyant and longest-
lasting of all pre-Columbian civilizations. Once humble village farmers, the Maya transformed their low-lying tropical homeland into a

landscape of great cities ruled by powerful lords. Between the last few centuries before Christ and A.D. 900, Classic Maya civilization flourished in the southern lowlands of Mexico, Guatemala, and Honduras. The collapse came suddenly. The great ceremonial centers of the Petén and the southern lowlands were abandoned and huge portions of the region were deserted, never to be reoccupied. In the city of Tikal alone, the population of over twenty-five thousand declined to just one-third of that. The survivors clustered in the ruins of the great masonry structures and tried to retain a semblance of their earlier life. Within a few generations, even they were gone.

The "Classic Maya collapse" is one of the great controversies of archaeology, but there is little doubt that droughts, fueled in part by El Niño, played an important role in the disaster.

Most of Mexico comes under the influence of the Intertropical Convergence Zone (ITCZ), which moves north during the Northern Hemisphere summer, bringing monsoonlike rain to the country between April and October. During the northern winter, the ITCZ moves south, toward the equator, and subtropical high pressure brings stable, dry conditions. The rainfall patterns vary constantly. When the monsoon is weak and suppressed and the ITCZ stays farther south than usual, prolonged droughts affect highlands and lowlands alike. This often occurs when the Southern Oscillation Index is low and El Niño conditions prevail over much of the tropical world.

The great El Niño of 1997–1998 parched much of Mexico, including ancient rain forest. Wildfires consumed about one-sixth of the Chimalapas rain and cloud forest in Chiapas, where at least fifteen hundred of the world's most endangered species live. Rare howler monkeys and panthers fled the flames, and rare orchids and lichens were incinerated. The spindly black trunks of thirty-meter trees rose like splinters from the charred land. Smoke drifted above the dense forest canopy like a delicate veil as the racing flames left thousands of trees brown-leafed and singed. Said Homero Aridjiis, Mexico's leading environmentalist: "We are losing an ecosystem that has taken thousands of years to form." Thousands of people fought over forty

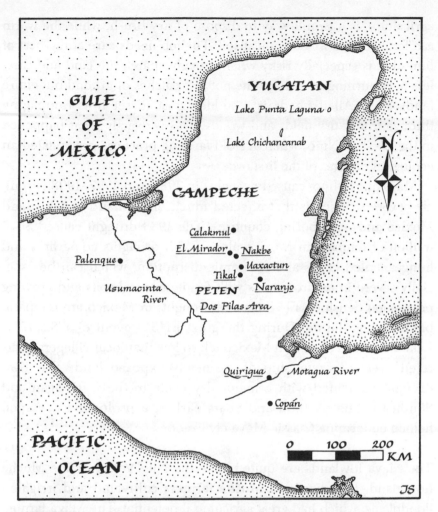

FIGURE 8.1   Archaeological sites and places mentioned in Chapter 8.

thousand hectares of brushfires: professional crews from Mexico and the United States, soldiers, townspeople, and Indian farmers. Much of the terrain was so steep and inaccessible that it took a four-day walk to reach a blaze.

Mexican farmers have set fires to clear their fields for thousands of years, since long before the brilliant Maya civilization flourished in Central America's lowlands. Fire setting, as old a practice as maize agriculture, is a simple way to dispose of dry brush, while the warm ash fertilizes the weak tropical soil. The job needs care at the best of times but is especially risky when droughts parch the land and careless field burning may consume not just patches of ground but entire landscapes. All it needs is a sudden gust to set fires rampaging through tinder-dry fields and woods. This is exactly what happened in the latest El Niño, when age-old farming practices precipitated an ecological disaster of the first order.

Maps of El Niño–caused droughts in 1983 and 1997–1998 chronicle severe dry spells that affected much of highland and lowland Mexico and neighboring countries. The 1983 drought caused $600 million of damage in Mexico alone. With droughts come fires and hunger, failed crops and economic disruption, which can be especially severe in environments with fragile tropical soils and growing population densities. El Niños and droughts in Mexico are inseparable and devastating. During the great ENSO event of 1782–1783, Lake Pátzcuaro in central Mexico fell so low that local villagers quarreled over the ownership of the newly exposed land. The 1793 drought coincided with a major dry cycle in the Caribbean and Southeast Asia. A thousand years earlier, a prolonged dry spell helped undermine Classic Maya civilization.

The Maya lowlands are quite unlike the Nile Valley or the Moche homeland. Egyptian and Moche farmers cultivated and irrigated floodplains, which had great agricultural potential. The Maya farmed the Petén-Yucatán peninsula that juts into the Gulf of Mexico, a vast limestone shelf lifted up from the depths of the ocean over an immensely long period of time. The porous limestone flattens out as one travels north into the Yucatán from the more rugged southern lowlands and looks just like a featureless green carpet when viewed from the air. The lowlands are hot and humid, with a six-month rainy season from May through October. Rainfall is less abundant and predictable than one might expect in such a low-lying tropical environment. Few permanent rivers except the Usumacinta and the Motagua provide navigable waters or abundant freshwater supplies. No annual inundation fertilizes the land with bountiful regularity. Most ancient Mayans obtained drinking water from circular sinkholes in the limestone, from swampy depressions, or from bottle-shaped cisterns with plastered rims that caught water during the wet season. Irregular rainfall plagued Maya farmers.

A dense forest once mantled the southern lowlands, where some of the greatest Maya cities flourished. Mahogany trees towered as high as forty-five meters above the ground, interspersed with sapodillas (now a source of chewing gum) and breadnuts. Abundant avocado and other fruit trees grew in the lower and middle layers of the forest. In places the woods gave way to patches of open savanna covered with coarse grass and stunted trees. As the centuries passed, the Maya cleared much of the primordial forest. What we see today is the regenerated vegetation of centuries.

The seeming uniformity of the lowlands is an illusion. The dense tree cover masks an astonishing diversity of local habitats, all of which presented special challenges to ancient Mayan farmers. Hot, humid, and unforgiving, the Maya homeland had few fertile soils, except in parts of the Petén and along larger river valleys. Pelting rain and intense tropical sunlight wreaked havoc with forest-cleared land, which soon became impossible to farm as a layer of brick-hard laterite formed on the surface. To cultivate such demanding fields by clearing and burning off the forest, then planting, required great experience and unlimited reservoirs of patience. A greater contrast with the Nile Valley is hard to imagine.

Like their modern descendants, the ancient Maya were plagued by droughts and decade-long climatic shifts, known to us from core borings into ancient lake beds. David Hodell, Jason Curtis, and Mark Brenner of the University of Florida cored the sediments of salty Lake Chichancanab in the Yucatán in a search for climatic data. Their core measured the changes in the oxygen-isotope ratio in shell carbonate preserved in the bottom sediment over many centuries. This, and the oxygen-gypsum ratio in the fine silt, allowed the scientists to reconstruct past changes in the ratio between evaporation and rainfall. They assumed that periods of drier climate were reflected by increased oxygen-isotope composition and a higher proportion of gypsum to calcite, the opposite being true in wetter cycles. Hodell and his colleagues developed a sequence of climatic change for the past eight thousand years, with an accuracy of about twenty years.

They found that Lake Chichancanab first filled in about 6200 B.C., at a time when Caribbean sea levels and the Yucatán's freshwater aquifer rose sharply. They continued to rise until about 4000 B.C. The core shows relatively wet conditions until about 1000 B.C., when conditions became drier and carbonate levels in the lake sediments increased as a result. Drying continued, peaking between about A.D. 800 and 1000, at the very time of the Maya collapse. The drought cycle of these two centuries was the driest period of the last eight thousand years.

The Chichancanab drought does not stand alone. Low lake levels in Lake Pátzcuaro, in Michoacán, central Mexico, and evidence of widespread brushfires in Costa Rica during the same two centuries also suggest widespread aridity. Hodell and his fellow researchers have also cored Lake Punta Láguna in Mexico's Quintana Roo, which loses most of its water through evaporation, meaning that changes in the oxygen 18 and 16 ratios are controlled almost entirely by the ratio between evaporation and rainfall. The lake sediments accumulated rapidly, so a core boring can produce a highly accurate, tree-ring-like chronology of rainfall changes. The Lake Punta Láguna sample revealed frequent and severe dry events between 165 B.C. and A.D. 1020. A particularly intense drought hit in A.D. 585, coinciding with the dry period recorded in the Quelccaya ice cap in the southern Andes that profoundly affected Moche civilization. This dry spell caused disruption in the Maya lowlands but was followed by a period of relatively abundant rainfall. Two centuries of vigorous growth and prosperity followed, and the Maya population grew rapidly. Like Chichancanab, Lake Punta Láguna witnessed a prolonged drier cycle between A.D. 725 and 1020, with two severe dry episodes in 862 and 986. The drought ended abruptly in 1020. Within a century the lake experienced some of the wettest conditions of the past eight thousand years. Since then, the climate has been marked by repeated wet and dry cycles that have lasted a decade or so, with some of the severest droughts unfolding during ENSO events.

The Yucatán lake cores display a long-term pattern of millennial shifts between abundant rainfall and drought and repeated multi-decade wet and dry cycles superimposed upon them. While some of these drought cycles coincide with ENSO episodes, twentieth-century observations in the Caribbean and Central America demonstrate a strong correlation between higher than average rainfall and unusually low pressure on the equator side of the high pressure near the Azores Islands. At the same time, the North Atlantic high and trade winds are displaced northward. The North Atlantic is unusually warm between ten and twenty degrees north, from Africa to the Americas. Enhanced convergence and cloudiness over the Caribbean and the adjacent low-latitude Atlantic region means much more moisture than usual. In contrast, dry years result from a much less intense annual cycle, and the opposite conditions over the North Atlantic. The droughts that afflicted the Maya in the eighth and ninth centuries resulted from complex, still little-understood atmosphere-ocean interactions, including El Niño events and major decadal shifts in the North Atlantic Oscillation, as well as two or three decade-long variations in rainfall over many centuries.

Despite these short-term climatic swings and a tropical homeland with fragile and only moderately fertile soils, Classic Maya civilization flourished for more than eight centuries before the harsh realities of overpopulation and environmental stress toppled its proud leaders. That it survived so long in such a demanding environment is a tribute to the skill of Maya farmers.

Like other tropical cultivators, the Maya used ancient slash-and-burn farming methods to grow maize and beans. Come late fall, the farmer would cut down a patch of forest on well-drained land during the end of the dry season, then burn off the wood and brush. This was a critical time in the farming year, when the air was thick with wood smoke and dust. Great clouds of gray smoke billowed into the washed-out blue sky as the afternoon wind blew fine ash and soot over everything. As the burn subsided, the ash and charcoal fell on the soil. The farmers and their families worked the natural fertilizer into the earth, then planted maize seed in holes poked into the soft ground with a stick. Timing was everything, for planting had to coincide with the first rain showers.

Such cleared gardens, called *milpa,* remain fertile for only about two years. The farmer must then move on to a new plot and begin again, leaving the original *milpa* to lie fallow for between four and seven years. When the Maya were purely village farmers, their settlements lay amid patchwork quilts of newly cleared plots and regenerating land, surrounded by thick forest that separated them from their neighbors. But as the farming population rose, expanding communities gradually ate up virgin land.

Slash-and-burn cultivation worked well enough when the Maya farming population was small, but the crop yields were never sufficient to support large settlements. Nor could the stocks of surplus grain feed more than a handful of nonfarmers, such as stone ax makers or priests. However, until the last few centuries before Christ, this simple farming system was the staple of an increasingly complex village society that flourished in a hot, low-lying environment with poor, shallow soils.

Could such an elemental farming system support an elaborate civilization of densely populated cities? Early-Maya archaeologists observed the never-ending cycle of clearance, planting, and harvest in modern times and assumed the same farming methods had indeed supported an ancient civilization of village farmers dispersed through the lowlands, who came together only to build elaborate ceremonial centers like Copán, Palenque, or Tikal, where few people actually lived. Wrote the Mayanist Sylvanus Morley in 1946: "The modern

Maya method of raising maize is the same as it has been for the past three thousand years—a simple process of felling the trees, of burning the dried trees and brush, of planting, and of changing the location of the cornfields every few years."[1]

Morley and his contemporaries knew little of tropical soils and believed the *milpa* lands were absolutely uniform over the entire lowlands when, in fact, the environment is exceptionally diverse. So they argued that *milpas* and nothing else had supported a basically rural society scattered widely over the lowlands. The Maya nation was a civilization without cities, ruled by peaceful religious leaders with a passion for calendars and the movements of the heavenly bodies.

Half a century later, we know Morley and his contemporaries were wrong. The Maya supported enormous cities and crowded landscapes with some of the most sophisticated agriculture ever practiced in the Americas.

In 1972 the geographer Albert Siemans and the archaeologist Dennis Puleston used aerial photographs to study extensive tracts of wetlands in Mexico's southern Campeche. They identified irregular grids of gray lines in ladder, lattice, and curvilinear patterns, which turned out to be long-forgotten raised field systems. Most of them were narrow, rectangular plots elevated above the low-lying, seasonally inundated land bordering rivers. These raised fields bore a close resemblance to the well-known *chinampas,* or swamp gardens, used centuries later by the Aztecs of highland Mexico. The Aztecs maintained thousands of acres of *chinampas* around their great capital, Tenochtitlán, which itself lay in the midst of a shallow lake. These highly productive fields produced several crops of beans and maize a year and supported more than 250,000 people in the Valley of Mexico alone.

The Siemans and Puleston research exploded the long-held "myth of the *milpa*" as the only source of Maya subsistence and seemed to explain how village farmers had managed to build flourishing cities. The Maya started draining and canalizing swamps at least two thou-

sand years ago, turning agriculturally useless land into highly productive acreage as long as there was ample groundwater. They also began cultivating steep hillsides.

Back in the late 1920s, a number of scientific travelers had noticed
extensive tracts of abandoned terracing on steep hillsides in the Maya
lowlands. But no archaeologists investigated these ancient field systems until the wetland research made them aware of the diversity of
Maya agriculture. The stone-walled terraces rise in serried rows,
thereby trapping silt that would otherwise cascade down dry hillsides
during torrential rainstorms. Like the heavily cultivated wetlands, the
terrace systems were a sign of intensive farming that made use of
every patch of fertile land.

Mayanists embraced the new evidence for intensive agriculture.
Here was an explanation for the seemingly mysterious population explosion that accompanied the first lowland cities: Faced with growing
numbers and a land shortage for *milpa* cultivation, the Maya turned
to swamp agriculture to support their burgeoning cities at about the
time the Romans gained mastery of the Mediterranean world.

The new paradigm turned the Maya from peaceful, priest-ruled
villagers into peace-loving, swamp-farming city folk who occupied
the entire lowlands. Some estimates placed the Maya population at
between eight and ten million people in A.D. 800, a staggeringly high
density for a tropical environment with such low natural carrying capacity.

There are no easy answers to how the Maya managed to support
so many nonfarmers. The 1970s and 1980s saw a shift in research
away from large-scale excavation of cities to the study of entire ancient landscapes and the hinterlands of Maya cities like Copán in
Honduras and Tikal in Guatemala. Researchers studied aerial photographs, cut long transects through clinging forest, and studied humble farmsteads and small villages in a search for changing Maya settlement patterns over many centuries. They soon found that by no
means all wetlands, with their irregular flood levels and water tables,
were capable of supporting high-yield raised fields. Some cultivators

terraced nearby hillsides. Other groups, a short distance away, never bothered. Furthermore, painstakingly compiled site distributions were sometimes at odds with the clusters of fertile soils, swampy ground, and other potentially farmable areas in the vicinity. What the Maya chose to farm in the past did not always match the best arable land today.

The more perceptive researchers now realized that they were looking at complex mosaics of landscapes that were farmed over many centuries. Over these many generations, the Maya's own perceptions of these landscapes and their potentials changed in radical ways. These altered discernments came from the institution of Maya kingship and from humanly induced and natural environmental change.

Maya kingship developed from deep roots in the past. Like all slash-and-burn farmers, the earliest Maya lived close to the land. Their survival depended on expert knowledge of soils and weather signs, animals, plants, and the subtle changes that marked the passage of seasons. The villagers peopled the forest with spirit beings, half animals, half humans. This was the world of the fierce jaguar, equally at home in water and in the trees. The multiple layers of the Maya cosmos began with the still, dark waters of the primordium, an eternity inhabited only by supernatural beings. Many centuries later the Quiche Maya *Popol Vuh*, an ancient Maya book of counsel, tells us: "The begetters are in the water, a glittering light. They are there, they are enclosed in quetzal feathers in blue-green."[2] The gods conferred and created an earth that emerged from the water, sprouting plants and people. From the very beginning, the living and spiritual worlds were one.

Long before civilization developed in the lowlands, the supernatural world surrounded Maya communities on every side. Their earth was the middle layer of a cosmos where *wacah chan,* the World Tree, linked the realms of heaven, the living, and the underworld. This powerful and very ancient ideology linked past, present, and future generations with imperishable bonds in a world where ancestors and gods validated the deeds of humans. From the beginning, shamans, those with the power to pass from the living to the supernatural worlds, assumed great importance in Maya society.

Village shamans were healers and priests, individuals with the ability to harness the power of hallucinogens and enter deep trances that allowed them to fly free in the cosmos. Gashing their penises with jagged stingray spines, they splattered blood that opened the gateway to the otherworld and passed effortlessly into the spiritual realm. Their trance performances became increasingly elaborate public ceremonies performed before ever larger audiences. By 300 B.C., the most able shamans had become powerful lords, and their hamlets had developed, first into small towns, then into imposing ceremonial centers that replicated the Maya world in stone and stucco.

In 150 B.C., growing cities like Nakbé and El Mirador in the Petén boasted imposing pyramids that bore plaster masks of gods and revered ancestors. El Mirador covered sixteen square kilometers and encompassed vast pyramid complexes. The Danta Pyramid, at the east end of the site, rose more than seventy meters from a natural hill. A network of carefully engineered causeways radiated out over swampy ground to nearby centers. The freshly plastered and painted temple facades bore testimony to a powerful new institution in Maya life: *Ch'ul Ahau,* sacred kingship. Maya cities became symbolic landscapes wrought in stucco and stone: The pyramids were sacred mountains, and carved stelae replicated trees in the forest. The temple was the mountain gateway to the all-important otherworld through which the shaman-lords passed on their trance-induced journeys into the spiritual realm.

The earlier Maya lords were anonymous. They left no written records, no glyphs recording their deeds or personal histories. Their only legacy to their successors was a repertoire of powerful rituals that validated the Maya world embedded in a cyclical calendar, the institution of kingship itself, and a distinctive architecture that replicated the cosmos. So powerful was this architecture that generations of great lords built new temples on the foundations of earlier ones.

Maya history and kingship were linked to the present and the other-world. Thus, the credibility of all Maya lords depended on their ancestry and their place in history.

About two thousand years ago, Maya lords began recording their deeds and genealogies with a distinctive and elaborate script. Today we can read more than two-thirds of Maya script, to the point that an archaeologist digging a temple can use the glyphs to guide the excavations and can date individual structures with the very inscriptions left by their builders. Maya glyphs tell us about astronomy and the passage of time. They also record much about kingship, grand historical events, and the rise and fall of dynasties. The rulers' own writings clothe silent pyramids and cities in political and religious garments. They tell us Maya society was a patchwork of competing city-states ruled by bloodthirsty lords, obsessed not with calendars and rituals but with genealogy and military conquest. Past masters of diplomacy and the manipulation of prestige, the ruthless Maya lords nurtured powerful ambitions that destroyed their environment and brought down their great cities.

The glyphs tell us that the king was state shaman, the intermediary with the otherworld, almost a form of family patriarch. Maya lords believed they had a divine covenant with the gods and ancestors. They depicted themselves as the World Tree, the conduit by which humans communicated with the supernatural realm. In Maya belief, trees were the living environment of human existence and a metaphor for royal power. So Maya rulers were a forest of symbolic World Trees within a natural forested landscape. Their stelae and inscriptions go to enormous pains to legitimize their descent from previous rulers and long-deceased parents.

Like the Egyptian pharaohs, Maya lords were divine rulers. The Maya worldview created serious and binding obligations between the lords and nobility and the thousands of farmers who fed them. The ruler's success in organizing agriculture and trade enriched the lives of everyone in spiritual and ceremonial ways. Each lord had serious responsibilities in organizing large-scale agricultural works and in gathering and distributing commodities of all kinds throughout his kingdom. A great ruler was far more than a secular ruler. He *was* Maya life, linked to his family and communities large and small by the institution of kingship and its genealogical ties to the founding ancestors. Unfortunately for the Maya, the form of kingship they chose depended on an agricultural economy and environment that could not sustain their rulers' insatiable demands.

The superficial resemblance to Egyptian kingship ends there, for Maya lords lived in a very different economic and political environment. Every Maya ruler, whether master of a huge city or a small town, lived in a world of shifting political alliances, ritual observances, and military campaigns undertaken to achieve economic goals or to capture prisoners for sacrifice. Diplomatic visitors passed constantly between one city and the next, to celebrate accessions, political marriages, or the formal designation of heirs. Most wars were between neighbors. One city would capture another. The ruler would be sacrificed by his conquerors. Many generations later the conqueror would become the conquered as the restless tides of political power ebbed and flowed.

The greatest royal dynasties endured for many centuries. Tikal's ruling family presided over the city from A.D. 219 until the ninth century. Tikal went from strength to strength. On January 16, 378, Lord Jaguar-Paw defeated and sacked his powerful neighbor Uaxactun, installing a vassal dynasty led by the successful general Smoking Frog. At the height of its powers, Tikal ruled over a kingdom of 2,500 square kilometers with an estimated population of more than 360,000 people. But the state was never secure: Rivals chipped away at its frontiers, and judicious political marriages changed delicately arranged diplomatic equations.

At least four major capitals flourished by the eighth century A.D.: Calakmul, Copán, Palenque, and Tikal. By this time Maya society was increasingly top-heavy with nobility, while agricultural productivity remained basically stable, since the farmers had exploited the limits of the environment. More than a quarter-century of excavations and archaeological surveys at Copán tell a haunting story of rapid growth and abrupt collapse.

Copán's pyramids and plazas covered twelve hectares, rising from the vast open spaces of the Great and Middle Plazas, where intricately carved stelae commemorate Copán's great lords. In A.D. 426, Lord Kinich Yax Kuk Mo (Sun-eyed Quetzal Macaw) arrived at Copán and founded a dynasty of fifteen rulers, who built temple after temple on the same location, known to archaeologists as the Acropolis. Over four centuries, Copán became a major kingdom of the Maya world. By the eighth century, more than ten thousand people lived in the Copán Valley close to the capital.

The story of Copán comes from archaeological surveys of the city's hinterland as much as of the urban core. Using aerial photographs and foot surveys, the survey teams have recorded more than 1,425 archaeological sites containing more than 4,500 structures over more than 135 square kilometers of the area surrounding the urban core. They have excavated a sample of 252 of the settlements and recovered many obsidian fragments, used by the Maya for ornaments, stone tools, and mirrors. Archaeologists value obsidian because they can measure the depth of the water absorption (hydration) layer on the surface and establish the time that has elapsed since the fragment was manufactured into a tool, or since a fresh edge was used. Over 2,300 such dates provide a chronology for startling changes in human settlement around Copán.

Between A.D. 550 and 700, the Copán state expanded rapidly. Most people lived within the urban precincts and immediately around the city. Only a small rural population dwelt in the surrounding countryside. From 700 to 850, the population increased by leaps and bounds to between 20,000 and 25,000 people. Eighty percent of the populace still lived within, or on the edge of, the city, at a time when the population was doubling every 80 to 100 years. However, there are signs of changing conditions. The rural population was expanding outward along the valley floor. That some villages began to cultivate less productive foothill areas some distance from town suggests that fertile agricultural land was in shorter supply. Meanwhile, Copán's burgeoning urban population became increasingly stratified. Eighty-two percent of the people lived in humble dwellings in a society that was becoming increasingly top-heavy.

Copán first ran into trouble in A.D. 738, when Lord Cauac Sky of the nearby vassal city of Quirigua turned on his master, captured him, and put him to death. For a short time the ruler became the ruled, but the great city was never destroyed. In 749, Lord Smoke Shell ascended to Copán's throne and embarked on an ambitious building program to validate his rule. His masterpiece was the magnificent Hieroglyphic Stairway, adorned with over 2,200 glyphs that celebrate the Maya lords' supernatural path. The stairway has the feel of a revivalist movement, an attempt to galvanize support behind the failing dynasty and a large nobility, riven by factionalism, that demanded all the expensive privileges of the elite.

Smoke Shell's son Yax-Pac (First Dawn) inherited a tottering kingdom. By this time the self-perpetuating Maya nobility was of unmanageable size, clustered in courtiers' residences close to the city. Yax-Pac took the unprecedented step of visiting high dignitaries' dwellings and dedicating their houses. We know this from surviving inscriptions that commemorate the lord's visit and his validation of the lineage held by each loyal courtier.

Even the ruler's prestige could not ward off Copán's collapse, for larger forces were at work. The hinterland surveys tell the story. As the dynasty ended in 820, rapid depopulation began. The urban core and periphery lost about half their population after 850, while the rural population increased by almost 20 percent. Small regional settlements replaced the great city and scattered villages of earlier times. By 1150, only five thousand to eight thousand people remained in the Copán Valley.

Copán was not alone. By A.D. 900, Maya civilization had collapsed over a wide area of the southern lowlands, in one of the most dramatic implosions of any early civilization in human history.[3]

Why did Maya civilization suddenly come apart? Everyone who studies the Classic Maya collapse agrees that it was brought on by a combination of ecological, political, and social factors. Patrick Culbert of the University of Arizona has shown that the population densities of the southern lowlands rose as high as two hundred people per square kilometer immediately before the collapse, over an area so large that it was impossible for people to adapt to bad times by moving to new, uncleared areas some distance away. The Maya had no fresh acreage left. Culbert believes the magnitude of the population loss after A.D. 800 was such that social malfunction alone cannot account for it.

By the time the collapse came, Maya agricultural production had approached its limits. Larger cities like Tikal may have transported large amounts of foodstuffs from one hundred kilometers away or more. This intensification of production had begun centuries earlier, when the Maya combined *milpa* cultivation with wetland field systems and terracing. In the short term, these intensification strategies worked at the local level, especially when the elite exercised increasing control over production and crop yields through tribute assessments, and perhaps through taking careful inventory of farmland. Some Mayanists believe the Maya eventually went a step further and attempted to standardize farming practices over wide areas.

Such an approach is feasible in environments like Egypt's Nile Valley or even in the river valleys along Peru's north coast, where extensive irrigation systems require close control and a degree of standardization to maximize the dissemination of precious floodwater. However, the Maya environment was highly diverse, with fragile soils and unpredictable rainfall. The farmers did not rely on mountain runoff but, even in the wetlands, on irregular seasonal floods and fluctuating water tables in the porous limestone. They were faced with escalating demands from their masters even as their environment deteriorated: Soils became exhausted, torrential rain falling on deforested lands caused extensive sheet erosion, and there were chronic shortages of firewood. The farmers were becoming desperate at the moment when a major drought cycle delivered the knockout punch.

When the great droughts of the eighth and ninth centuries came, Maya civilization everywhere was under increasing stress. As we saw at Copán, the nobility was top-heavy and ridden with factionalism. Tikal's leaders were importing food from far away. Maya lords were demanding more and more from their subjects at a time of frenzied competition and military activity. They may have tried to standardize agricultural methods to increase crop yields, a strategy that could never work in such a diverse environment where the soils were already near exhaustion.

The drought was the final straw. Crop yields fell catastrophically, and tribute payments dwindled rapidly. Perhaps ambitious nobles took advantage of troubled times to further their interests as well-established dynasties like that of Copán saw their spiritual and political authority withering in the face of a restless and hungry populace. Unlike the Egyptian pharaohs, the Maya had no strong provincial governors to take drastic steps to conserve floodwater. Indeed, the environmental damage was beyond the capability of any leader to repair. Thousands starved as the survivors turned on their leaders and abandoned once-prosperous cities. Meanwhile, the village farmers in the countryside dispersed into smaller communities of a size that enabled them to support their families and kin without the burden of tribute payments in a world where the spiritual authority of many centuries was now bankrupt.

The collapse did not come without turmoil and war. The archaeologist Arthur Demarest has excavated a once-prosperous city named Dos Pilas in northern Guatemala, founded by a renegade noble from Tikal in A.D. 645. Dos Pilas's later rulers embarked on ambitious military campaigns that enlarged their territory to nearly four thousand square kilometers by the mid-eighth century. Dos Pilas became a wealthy trade center ruled by powerful lords who boasted on their monuments of their diplomatic and military skills. By 761, they had overextended themselves. Nearby Tamrindo attacked its once-powerful neighbor and killed the ruler. As the invaders built rough defensive walls, the surviving nobles fled and built a new fortified center at Aguateca, atop a craggy hill surrounded on three sides by steep cliffs. Attacked repeatedly, they held on for another half-century before intensive warfare drove the people into scattered towns and villages, where they fortified even their fields. With only defended acreage to farm, the farmers' crop yields must have fallen dramatically. In a last desperate stand, the surviving Aguatecans created an island fortress on a peninsula in Lake Petexbatun by digging three moats across the neck of land, one over 140 meters long. Even this outpost perished. The inhabitants abandoned it in the 800s.

The Maya collapse is a cautionary tale in the dangers of using technology and people power to expand the carrying capacity of tropical environments. The first time I visited the ruins of Tikal, I marveled at the size and complexity of the pyramids and plazas, built by a people living in a humid rain forest with the most fragile of tropical soils. Tikal is a staggering human achievement, built at an enormous environmental price. The city's royal dynasty lived on borrowed time, importing food, exploiting their subjects, and exhausting the land beyond recovery. Thousands of people dwelt on a landscape that could support but a few souls per square kilometer. A burgeoning population flourished and grew as long as there was ample rainfall and room to expand. Even then, the danger signs were present: reduced crop yields, perhaps widespread malnutrition, and social unrest fostered by a rigid and unyielding social order. The droughts were the knock-out blow that undermined the authority of divine lords.

The effects of the long dry spells rippled across the Maya world. Southern lowland civilization collapsed almost completely as the center of Maya life shifted into the northern Yucatán. There the water table was closer to the surface, so cities in the northern lowlands could survive by tapping lakes and natural reservoirs. In the south, where the water table is much lower, the cities relied on reservoirs, rainwater cisterns, and artificial *aguadas* for water supplies. When these dried up, large communities were in deep trouble. The same dry conditions pummeled crop yields just as the cumulative effects of centuries of overexploitation of fragile and marginal soils, widespread deforestation, and sheet erosion took their toll. When it did rain, the torrential downpours swept away unprotected soil. Kingship faltered in the face of drought and restless, hungry commoners. The elaborate superstructure of city-states collapsed like a stack of cards. Some of the greatest Maya cities dissolved in violence and social revolution within a few generations.

Atmospheric circulation changes far from the Maya homeland delivered the coup de grâce to rulers no longer able to control their own destinies because they had exhausted their environmental options in an endless quest for power and prestige. The survivors of the disaster did what the Anasazi Indians of the American Southwest did two centuries later during another catastrophic drought. They dispersed into small, self-sustaining villages where their descendants live to this day.

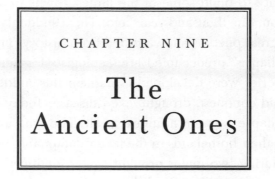

Survival, I know how this way.
This way, I know.
It rains.
Mountains and canyons and plants grow.
We traveled this way, gauged our distance by stories
*   and loved our children . . .*
We told ourselves over and over again,
We shall survive this way.
*   —Simon Ortiz of Acoma Pueblo*

Chaco Canyon can be a haunting place when the setting sun bathes its reddish cliffs in a rosy light. The shadows lengthen as the weathered masonry of the Anasazi great houses merges into darkness. Five thousand people once lived in this canyon. I stop and look down into the depths of the great kiva at Casa Rinconada, imagining the clinging wood smoke, the pine torches, the tumult of the masked dancers moving back and forth. A lizard scuttles across the earthen floor and vanishes into the *sipapu,* the sacred way that brought humans into the living world. The dancers are long gone, for drought dispersed the Chaco Anasazi nine centuries ago.

The Anasazi, the "ancient ones" of the Southwest, rose from complete obscurity to build some of the largest towns in ancient North America about one thousand years ago. Then, suddenly, they moved out of their great pueblos and vanished from history. The mysterious "Anasazi collapse" preoccupied archaeologists for generations. All kinds of theories were developed to explain this sudden disappearance: war and conquest, drought, even disease. Today we know the so-called collapse was not a failure at all. The Anasazi simply dispersed from their homelands in the face of untenable environmental conditions during two major drought cycles identified from ancient tree rings: A.D. 1130–1180 and the "great drought" of A.D. 1275–1299.

This tale of climate change, drought, and, above all, continual movement has been reconstructed from archaeology, cutting-edge science, and Pueblo Indian perspectives on human existence. The secret of the Anasazi's response to climate change lies in their complex, but little-known, ideology, in which the notion of movement played an important part.

I could not live off the land in the arid Southwest for a week, let alone years. The Anasazi survived and flourished in a diverse environment of desert, plateau, and mountains for many centuries because they knew their homeland intimately. They inherited the cumulative knowledge of ten thousand years of foraging in an unrelenting landscape, where survival depended on a close knowledge of alternative foods in times of hunger. The Anasazi added farming to this expertise, for they learned how to predict where cultivation would be successful. Above all, they were experts at water management.

The Colorado Plateau of the northern Southwest was the core of Anasazi country. Most of their homeland was between 1,200 and 2,400 meters above sea level, with rapid changes in elevation and accompanying environmental diversity. Like the distant Andes Mountains, this is a vertical environment, where for every 330-meter rise in elevation, the temperature falls about two and a half degrees Celsius.

Rainfall increases with height above sea level, but its distribution depends on the prevailing winds, distance from mountains, and slope direction. The greatest diversity of edible wild plants and animals occurred between 1,400 and 2,000 meters above sea level. Corn can be grown up to 2,130 meters but requires irrigation below 1,700 meters. The Anasazi farmed maize, beans, and squash with great success, but they always relied on wild plant foods and were always ready to move, since the fertility of their environment varied dramatically from one area to the next.

El Niño's influence is weaker in Anasazi country, but rainfall is all-important. Their homeland in the northern Southwest lies between two major sources of moisture: the Gulf of Mexico to the south, and the Pacific Ocean to the west, both of which are influenced strongly by ENSO episodes. Summer rain, which falls between July and September, comes from both areas, while winter precipitation derives from the large-scale frontal systems that originate on the California coast and sweep across the region between December and March. Fall and spring are dry, but the Anasazi were vulnerable to short-term rainfall shifts caused by changing atmospheric conditions in the Pacific and Gulf of Mexico. When the Southern Oscillation generated a strong El Niño, winter rainfall could increase considerably, while summer precipitation remained relatively stable. In other years variations in winter rainfall from the Pacific caused by northward shifts of the storm track brought irregular and sometimes prolonged drought cycles to Anasazi lands. We sometimes see these cycles reflected in widespread traces of fires in tree-ring records. We still do not know what distant interactions of the atmosphere and the ocean caused great fluctuations in southwestern rainfall.

Anyone farming this environment knew the value of water conservation and the importance of mobility.

"We come from some place under the earth. When they came out from there they started coming and moving and then they settled and they stood up again and then they started moving again. When they started moving they started moving to another location and then they stayed there for a few years." Stories of the past told by Pueblo Indians like the Tewa of northern New Mexico share one common element—movement. Movement is one of the fundamental ideological concepts of Pueblo thought, because mobility perpetuates human life. Says the Tewa Tessie Naranjo: "Movement, clouds, wind, and rain are one. Movement must be emulated by the people."[1] Keeping on the move was the way to survive.

Pueblo history has always been a continual process of moving, settling down, then moving again. People have moved, joined together, and separated again since long before farming began in the Southwest over three thousand years ago. The Pueblo believe that without movement, there is no life. Their Anasazi predecessors must have believed the same thing. They could not otherwise have survived in one of the harshest farming environments in the Americas.

A land of unpredictable rainfall and frequent droughts cannot support more than a handful of people on one hundred square kilometers. The remotest forager ancestors of the Pueblo Indians developed great expertise with wild plant foods, the realities of sparse water supplies, and the habits of small game. They survived, even flourished, in an inhospitable and highly diverse environment for thousands of years, responding to drought and long-term climatic changes by falling back on less nutritious foods and, above all, by moving across the land and fostering close kin ties. The habit of constant adjustment to changing environmental conditions was deeply ingrained in the southwestern farmer's psyche from the very beginning.

The ancestors of the Anasazi were already farming the valleys and hillsides of the Colorado Plateau when Rome was still a growing town and the Assyrians held sway over much of southwestern Asia. But their successful adaptation to their homeland was never in long-term equilibrium. Changing rainfall patterns, constant ebbs and flows of farming populations, and numerous behavioral changes danced an intricate waltz through the centuries. Fundamental to the dance was carrying capacity: the ability of the land to support an ever changing number of people per square kilometer.

Thanks to a gifted astronomer with an interest in sunspots, we have an extraordinary understanding of the demanding Anasazi world of one thousand years ago.

Andrew Ellicott Douglass was an astronomer at the Lowell Observatory in Flagstaff, Arizona, and an expert on the relationships between solar activity and climate on earth. In 1901, needing weather data to compare to the known twenty-two-year sunspot cycle, he turned to the annual growth rings of trees as a possible source of climatic records. When his initial results were somewhat unpromising, he almost gave up the work, but he happened to mention the subject during a lecture in New York. By chance, the southwestern archaeologist Clark Wissler was in the audience, and he was so impressed with the potential of tree-ring dating (often called dendrochronology, after the Greek word for tree, *dendros*) that he made contact with Douglass and persuaded him to resume his experiments.

In 1914 Douglass established a quantitative relationship between tree growth and climate when he demonstrated the strong correlation between ring width and the rainfall of the previous winter. Soon he was able to date segments of trees by comparing the inside sections of logs and the sets of rings on them to an expanding master curve of tree rings that began with modern logs and extended back centuries into the past. By 1918 Douglass was using a special core borer to sample ancient pueblo beams from Anasazi structures and old Spanish churches. It took another decade for him to bridge the gap between historical tree rings and his "floating chronology" from Anasazi pueblos like Pueblo Bonito in Chaco Canyon. The missing rings came from a large pueblo lying under a Mormon barnyard at Show Low, Arizona, where a charred log bore rings that dated back to A.D. 1237 and linked the two ring sequences.

Throughout his early experiments, Andrew Douglass insisted on rigorous procedures that set the standards for modern tree-ring science. He and his contemporaries made all their comparisons by eye and with hand calculations, but their accuracy set the stage for today's computer-based analyses that allow the simultaneous analysis of scores of tree-ring and environmental variables over many centuries. These procedures have established consistent relationships between tree growth and environmental factors, allowing us to reconstruct climatic change during Anasazi times with remarkable accuracy. Nevertheless, modern tree-ring experts like Jeffrey Dean of the University of Arizona can date a log segment just by looking at it. Douglass's methods still lie at the core of the science.

Dendrochronologies for the Anasazi are now accurate to within a year, giving us the most precise time scale for any early human society anywhere. In recent years the Laboratory of Tree-Ring Research at the University of Arizona has undertaken a massive dendroclimatic study that has yielded a reconstruction of relative climatic variability in the Southwest from A.D. 680 to 1970. The same scientists, headed by Jeffrey Dean, are now producing the first quantitative reconstructions of annual and seasonal rainfall, as well as of temperature, drought, and stream flow for the region. Such research is extremely complex, involving not only tree-ring sequences but intricate mathematical expressions of the relationships between tree growth and such variables as rainfall, temperature, and crop yields. These calculations yield statistical estimations of the fluctuations in these variables on an annual and seasonal basis.

By using a spatial grid of twenty-seven long tree-ring sequences from throughout the Southwest, Dean and his colleagues have compiled maps that plot the different station values and their fluctuations like contour maps, one for each decade. These maps enable them to study such phenomena as the progress of what Dean sometimes calls the "great drought" of A.D. 1276–1299 from northwest to southeast across the region. In 1276 the beginnings of the drought appeared as negative standard deviations from average rainfall in the northwest, while the remainder of the region enjoyed above-average rainfall. Over the next ten years, very dry conditions expanded over the entire Southwest before improved rainfall arrived after 1299. This form of mapping allows close correlation of vacated large and small pueblos with short-term climatic fluctuations.

When the research team looked at the entire period from A.D. 966 to 1988, they found that the tree-ring stations in the northwestern region accounted for no less than 60 percent of the rainfall variance. In contrast, stations in the southeastern part of the Southwest accounted for only 10 percent. This general configuration, which persisted for centuries, coincides with the modern distribution of seasonal rainfall in the Southwest: Predictable summer rainfall dominates the southeastern areas, while the northwestern region receives both winter and summer precipitation. Winter rains are much more uncertain. When the scientists examined this general rainfall pattern at one-hundred-year intervals from 539 to 1988, they observed that it persisted most of the time, even though the boundary between the two zones moved backward and forward slightly.

But this long-term pattern broke down completely from A.D. 1250 to 1450, when a totally aberrant pattern prevailed in the northwestern area. The southeastern region remained stable, but there was major disruption elsewhere. For nearly two centuries the relatively simple long-term pattern of summer and winter rains gave way to complex, unpredictable precipitation and severe droughts, especially on the Colorado Plateau. The change to an unstable pattern would have had a severe impact on Anasazi farmers, especially since it coincided with the great drought of A.D. 1276–1299.

Why did this breakdown occur? Dean divides the relationship between climatic change and human behavior into three broad categories. Certain obvious stable elements in the Anasazi environment have not changed over the past two thousand years, such as bedrock geology and climate type. Then there are low-frequency environmental changes—those that occur on cycles longer than a human generation of twenty-five years. Few people witnessed these changes during their lifetimes. Changes in hydrological conditions, such as cycles of erosion and deposition along stream courses, fluctuations in water table levels in river floodplains, and changes in plant distributions, transcend generations, but they could affect the environment drastically, especially in drought cycles.

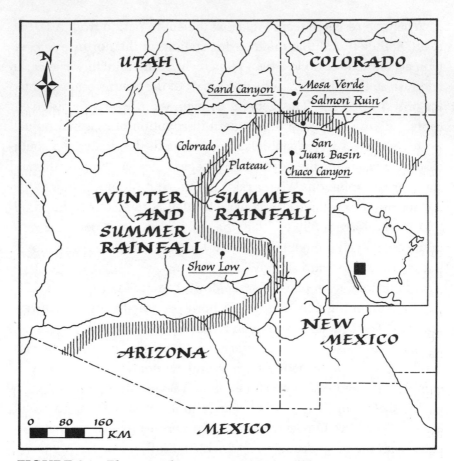

FIGURE 9.1 Places and sites mentioned in Chapter 9, also areas of rainfall in the North American Southwest. The shaded line marks the frontier between the winter and summer rainfall patterns of the northwestern region and the more predictable summer rainfall of the southeastern area.

Shorter-term, high-frequency changes were readily apparent to every Anasazi: year-to-year rainfall shifts, decade-long drought cycles, seasonal changes, and so on. Over the centuries the Anasazi were probably barely aware of long-term change, for the present generation and their ancestors enjoyed the same basic adaptation, which one could call a form of "stability." Cycles of drought, El Niño rains, and other high-frequency changes required temporary and flexible adjustments, such as farming more land, relying more heavily on wild plant foods, and, above all, moving across the terrain.

Such strategies worked well for centuries, as long as the Anasazi farmed their land at well below its carrying capacity. When the population increased to near carrying capacity, however, as it did at Chaco Canyon in the twelfth century, people became increasingly vulnerable to brief events like El Niños or droughts, which could stretch the supportive capacity of a local environment within months, even weeks. Their vulnerability was even more extreme when long-term changes—such as half a century or more of much drier conditions—descended on farmland already pushed to its carrying limits. Under these circumstances, a yearlong drought or torrential rains could quickly destroy a local population's ability to support itself. Like the Maya, Anasazi households in this situation had few options left but to disperse. The Moche did not have even that alternative.

The Anasazi lived and farmed as households. They could live with less risk that way, for they could grow enough food for themselves and for kin in nearby villages in times of hunger. They adapted to their harsh surroundings by developing an expertise at selecting soils with the correct moisture-retaining properties on north- and east-facing slopes that received little direct sunlight. Their farmers planted on river floodplains and at arroyo mouths, where the soil was naturally irrigated. They diverted water from streams and springs, using every drop of rainfall runoff they could. They developed maize with distinctive root structures that could be planted deep in the soil, where retained ground moisture would nourish the growing plants. Everything was done to reduce the chance of crop failure. As a matter of routine, the cultivators dispersed their gardens widely over the landscape to minimize the risk of local drought or flood. They learned how to shorten the growing season from the usual 130–140 days to sometimes closer to 120 days by planting on shaded slopes, at varying elevations, and in different soils. As long as they stayed within the carrying capacity of their environment, they had plenty of options.

By the early 800s, some Anasazi lived in larger settlements. Small hamlets became clusters of rooms and storehouses, built in contiguous blocks that formed much bigger communities than the hamlets of earlier times. Many Anasazi remained in scattered villages, but much denser populations flourished in places like Chaco Canyon and to the north at higher elevations in the Moctezuma Valley and Mesa Verde area of the Four Corners region. Between A.D. 840 and 860, when Classic Maya civilization was at its height, some Anasazi communities in the Dolores Valley area housed dozens of families. We do not know why so many households moved into closer juxtaposition, but tree rings tell us rainfall was better than average during the early ninth century.

Herein lies the story of constant movement. Anasazi life was never completely stable, for the vagaries of dry and wet years kept households and communities in a state of constant adjustment. It is no coincidence that modern-day Pueblo Indians have complex social dualities, among them one between winter and summer people. The winter people are hunters and gatherers of wild plants, and the summer folk are the farmers. This does not mean that winter people never cultivate the land, simply that their contribution is to supply game and plant foods to society as a whole. We can be sure that the Anasazi had many forms of duality, for fission and factionalism are constant in a society where the household and ties of kin are all-important as ways of coping with drought and adjusting to an unpredictable homeland.

In the mid-ninth century, the Anasazi in the Four Corners region left their large pueblos and dispersed widely over the landscape in response to a cycle of uncertain rainfall. However, their Chaco relatives to the south lived in a drier environment where the best-watered place for kilometers around was their home canyon. In a sense, they were trapped within its cliffs, so they had to develop new strategies for coping with variable rainfall.

Chaco Canyon, New Mexico, lies in a stark, dramatic landscape. The precipitous cliffs of the deep canyon glow yellow-gold in the setting sun and contrast with the softer tones of desert sand, sage, and occasional cottonwood trees. A few springs and the water that seeps from the canyon walls make this an attractive place for maize agriculture. Small lakes just south of Chaco attracted Anasazi farmers long before any pueblos rose in the shadow of its cliffs.

More than twenty-four hundred archaeological sites tell the story of this place. In about A.D. 490, the first permanent villages appeared in the south of the canyon as small groups of the local Anasazi took advantage of locally developed corn strains and more plentiful rainfall. The farmers dwelt in tiny hamlets. By 750, many of the villages had become small pueblos, clusters of rooms shaped into small arcs that faced southeast to take advantage of the winter sun. The arc shape made each household equidistant from the kivas, the sacred chambers in the center of the pueblo.

During the ninth and tenth centuries, the period of the Maya collapse, summer rainfall was highly variable. Instead of dispersing, the Chaco Anasazi built three "great houses," large pueblos located at the junctions of major drainages. The largest of these, Pueblo Bonito near the northeast wall of the canyon, stood five stories high along its rear wall and remained in use for more than two centuries. The first pueblo structure rose on the site of a pithouse village and expanded rapidly. Construction of the semicircular town began in the 850s and accelerated between 1030 and 1079, when the builders added the two great kivas. In its eleventh-century heyday, Pueblo Bonito had at least six hundred rooms in use and could house about one thousand people. No one knows why the Chaco people congregated in such enormous pueblos. It may have been a way to avoid building on valuable agricultural land and to concentrate food resources. Interestingly, by no means everyone lived in the great houses, which lie on the north side of the canyon. The opposite wall sheltered smaller settlements, many of them little more than a few households, located in areas where farmland was more spread out.

Archaeologists disagree as to whether Pueblo Bonito and the other great houses were built according to a master plan. The town was really an agglomeration of households, organized along kin lines; each kin group built its own rooms inside the semicircular outer wall, which was erected by communal effort. The outer walls were battered to support the massive weight of five stories of rooms, which were terraced to allow access without an interior system of ladders. Each kin group used its smaller kivas in the heart of the room blocks as workshops and as places for the education of children, storytelling, and family ceremonies. The great kivas, twelve to fifteen meters across, stood on either side of a line of rooms that divided the pueblo into two areas that reflected the duality of social organization in Anasazi pueblos, like the winter and summer people of modern pueblos. Here the people gathered for more formal ceremonies and to make decisions about the governing of the community as a whole.

By A.D. 1050, five great pueblos dominated Chaco Canyon. We do not know how many people lived within its cliffs and in the immediate vicinity. Estimates range from two thousand to as high as twenty thousand. The archaeologist Gwinn Vivian has calculated the potential carrying capacity of the canyon soils. He believes that no more than about fifty-five hundred people ever lived in the canyon: The land could support no more. Vivian is probably right, for Chaco was no Garden of Eden. The nearby mesa is the boldest topographic feature for many kilometers and a rich source of wild plant foods. Despite this diversity of resources, the Chaco people were trapped inside the narrow confines of their canyon. Unable to disperse easily over a wider territory, they took advantage of long-standing kin and trade links with communities living elsewhere on the Colorado Plateau and made themselves the hub of a much wider world.

Eleventh-century Chaco became a major center for processing turquoise into finished ornaments. Chaco itself had no turquoise, but the people had access to, and may have controlled, sources near Santa Fe, 160 kilometers to the east. We know this because more than sixty thousand turquoise fragments, many of them partially fabricated ornaments, have come from Chaco villages and pueblos.

Trade was only part of the picture, for Chaco seems to have exported households as well. By A.D. 1115, at least seventy communities
had dispersed over more than sixty-five thousand square kilometers
of northwest New Mexico, and parts of southern Colorado enjoyed
economic, political, and ritual links with Chaco's great houses. When
the archaeologist Cynthia Irwin-Williams investigated the Salmon
Ruins near Bloomfield, New Mexico, she uncovered a 290-room
pueblo with a great kiva and a tower kiva, all of which was carefully
laid out before a room was built. The inhabitants constructed the
pueblo in three orchestrated stages. Irwin-Williams found large quantities of Chaco-style pottery in the tower kiva and rooms nearby, suggesting a strong link between the religious beliefs practiced at Chaco
and those at Salmon.

Places like Salmon were not colonies but rather descendant communities, places established near good agricultural land as a result of
population growth or factionalism within older pueblos. Some were
founded to exploit nearby resources such as turquoise. In most
places, the people used Pueblo Bonito–style architecture, often
choosing dramatic settings such as cliffs or canyon walls for their
pueblo. It is as if the environment were imbued with mythic qualities,
so the placing of pueblos came from a powerful but now-forgotten sacred geography that shaped the Anasazi world.

If we are right in thinking that places like Salmon were founded by
Chaco households, or that they absorbed excess population, then the
extraordinary road system built by the Chacoans makes sense. Chacoan "roads" were first identified in the 1890s and again in the 1920s.
In the 1930s, early aerial photographs revealed faint traces of what
appeared to be canals emanating from the canyon. During the 1970s
and 1980s, a new generation of investigators used aerial photographs
and side-scan radar to place the canyon at the center of a vast ancient
landscape. Over 650 kilometers of unpaved ancient trackways link
Chaco in an intricate web with over thirty outlying settlements. The

roads are up to twelve meters wide and were either cut a few centimeters into the soil or marked by low banks or stone walls. They run straight for long distances, in one instance as long as ninety-five kilometers. Each approaches the canyon, then descends via stonecut steps down the cliffs to the valley floor. There they merge in the narrow defiles and split, each leading to a different great house. At three locations where they merge there is a groove in the center of the road that clearly demarcates one side from the other.

The Chaco road system has been depicted as a network of trackways that brought hundreds or even thousands of traders and pilgrims to the canyon for major rituals. One model sees Chaco as a place of pilgrimage housing that maintained a skeletal population for most of the year. At the solstices and other times, people from the outlying communities flocked to Chaco to trade and celebrate major rituals. They stayed in the great houses, whose storerooms provided food for all the visitors. This kind of argument assumes that the Chaco Anasazi lived in a hierarchical society presided over by powerful chiefs, who controlled both trade and the fabric of ritual life for their widely scattered communities.

This ingenious theory stumbles on two points. First, the focus of Anasazi life was the household and ties of kin, which were the means by which people fed themselves and passed on accumulated expertise about making a living from one generation to the next. Second, many of the Chaco roads go nowhere, although they are linked to a great house or kiva. We Westerners tend to think of roads as traveling from A to B, and we have thus tended to join incomplete segments of Chacoan roads with straight dotted lines. The roads may not have actually been joined. While major north and south tracks radiated from Chaco, only about 250 kilometers of the roads have been verified on the ground.

A more likely explanation lies in Pueblo cosmology. The so-called Great North Road travels sixty-three kilometers north from Chaco before it disappears abruptly in Kutz Canyon. North is the primary direction among Keresan-speaking Pueblo peoples, who may have ancestry among the Chaco people. North led to the origin, the place where the spirits of the dead traveled. Perhaps the Great North Road was an umbilical cord to the underworld and a conduit of spiritual power. The Keresan also believe in a Middle Place, a point where the four cardinal directions converge. Pueblo Bonito is laid out according to these directions and may have served as Chaco's Middle Place.

Thus, Chaco and its trackways may have formed a sacred landscape that gave order to the world and linked outlying communities with a powerful Middle Place through spiritual ties that remained even as many households moved away from the canyon. Think of a giant ideological spider's web with a lattice of obligations among its component parts, and you probably have a credible model for Chaco's role in the eleventh-century Anasazi world. The great houses of the canyon lay at the center of this web of interconnections (named the "Chaco phenomenon" by Cynthia Irwin-Williams) with communities many kilometers away through kin ties and regular exchanges of food and other commodities. Gwinn Vivian calls the landscape a powerful statement, which he articulates as "We the Chaco."

Between A.D. 1050 and 1100, the rains were plentiful. Building activity engulfed the canyon as the great houses expanded. The Chaco phenomenon prospered, perhaps longer than it might have in drier times. The steady rise in the Chaco population was not a serious problem as long as winter rainfall fertilized the fields. Then, in 1130, fifty years of intense drought settled over the Colorado Plateau. "We the Chaco" became a meaningless fiction in the face of crop failure and famine. Soon the outlying communities ceased to trade and share food with the great houses, forcing the canyon towns to rely on their own already overstressed environment. The strategy they had adopted centuries earlier fell apart and trapped them in their home under more desperate conditions than ever before. The only recourse was deeply ingrained in Anasazi philosophy—movement. Within a few generations, the great pueblos stood empty; well over half of Chaco's population had dispersed into villages, hamlets, and pueblos

far from the great arroyo. Those who remained settled in small communities that even the parched land could support. But they were
gone by the early 1200s.

We have, of course, no means of knowing how many people perished when the fifty-year drought desiccated the land. It was but a
short-term episode on a much larger climatic canvas that brought
long-term fluctuations in rainfall patterns, water tables, and river
flows at about five-hundred-year intervals. The emptying of big
houses like Pueblo Bonito and Chetro Ketl seems like an epochal
event at a distance of nine hundred years, but it was merely part of
the constant ebb and flow of Anasazi existence.

Some Chaco Anasazi moved northward to the northern San Juan
Basin, where a magnificent flowering of ancestral Pueblo culture endured for another century. The Four Corners region is higher and
better watered, with many valleys and canyons where dry farming
worked well and ample game and wild plants could act as a cushion
during hungry months.

During the twelfth century, hundreds of Anasazi households
moved from dispersed communities into large towns, situated by the
banks of rivers, in sheltered valleys, and built into natural rock shelters in the walls of deep canyons. Some large settlements, like Sand
Canyon Pueblo in the Four Corners area, boasted about seven hundred inhabitants. Sand Canyon surrounded a large spring, as did
other large towns in the area. Around A.D. 1250, the residents-to-be
erected a huge enclosure wall, which may have taken thirty to forty
people two months to build. Over the next thirty years, they added
over 20 separate room blocks, which incorporated at least 90 kivas
and about 420 rooms. However, each household maintained its own
identity, with its own cluster of structures—a living space, a storage
room, a place to eat, and a kiva—just as it had done in its original dispersed settlement. At the same time, multiple households dwelt
within a single architectural complex, as if the wider ties of the family
had become more important than in earlier times.

Sand Canyon and other open-country pueblos in the Moctezuma Valley area achieved considerable size, but none are as famous or as spectacular as the Cliff House in Mesa Verde. In 1888, the celebrated cattle rancher and pueblo explorer Richard Wetherill was the first Westerner to visit the Cliff House. Three years later the Swedish scientist Gustav Nordenskiöld rode through monotonous piñon forest for hours until he suddenly emerged on the edge of a precipitous canyon. Across the valley he saw the Cliff House, "framed in the massive vault of rock above and in a bed of sunlit cedar and piñon trees below. . . . With its round towers and high walls rising out of the heaps of stones deep in the mysterious twilight of the cavern and defying in their sheltered site the ravages of time, it resembled at a distance an enchanted castle."[2]

Cliff House, with its 220 masonry rooms and 23 kivas, has a spectacular setting but actually differs little from large pueblos elsewhere. Perhaps half the population of nearby Chapin Mesa occupied the site. Nordenskiöld was the first to attempt a scientific excavation in the pueblo. The fine dust in the rooms blew everywhere, but the dry conditions preserved their contents in perfect condition: Fragments of cotton cloth, wooden digging sticks, lengths of string, and hide fragments lay everywhere, as if the Anasazi inhabitants had been gone for only a day.

The Mesa Verde Anasazi lived at more than twenty-one hundred meters above sea level, in mountainous terrain where winter temperatures can reach minus twenty degrees Celsius. However, only a small number of them dwelt in inaccessible canyons. Many more farmers inhabited the large drainages northwest of Mesa Verde, where population densities rose rapidly—from 13 to 30 people per square kilometer during the tenth century to as much as 130 three centuries later. At the same time, average village size doubled from six to twelve rooms. Soon a growing population had reached the limits of the carrying capacity of the land. Disaster was just over the horizon.

A team of scientists has studied agricultural productivity in the Sand Canyon region, relying both on modern-day environmental and agricultural data and on tree-ring readings that provide a measure of ancient conditions. The environmental scientist Carla Van West has reconstructed the severity of droughts for the month of June between A.D. 900 and 1300 for soils at five different elevations. She also calibrated variations in moisture levels with potential agricultural productivity, using data from crop fields between 1931 and 1960.

Van West put all her data into a vast geographical information system that allowed her to create environmental models, display contour maps, then overlay them with areas of potential agricultural productivity for a specific year. She ended up with a graph of potential maize production in the study area over the period when the Anasazi occupied and then abandoned the Sand Canyon area. Using modern land carrying capacity data, Van West estimated that the area could have produced enough maize to support an average local population of about 31,360 people at a density of 21 people per square kilometer over a four-hundred-year period. Her figures show that the twelfth-century drought that caused the Anasazi dispersal at Chaco had little effect in regions like Sand Canyon where there was still enough land to support the dispersed farming population. In other words, the people had enough room to make the movement that was central to their survival.

Van West was also able to show that potential agricultural productivity varied considerably from place to place and from year to year. The farmers tended to locate near consistently productive soils. They could survive the harshest of drought cycles *if* there were no restrictions on mobility or on access to the best soils, and *if* they could acquire food from neighbors when crops failed. However, their ability to move was severely restricted once population densities approached the carrying capacity of the land and the people had cultivated effectively all the most productive soils. At that point, surviving extreme short-term climatic change was much harder, especially when longer-term climatic cycles happened to coincide with a serious drought cycle, as happened during the great drought of A.D. 1276–1299.

As the drought cycle hit, pueblo construction suddenly slowed; it had ceased altogether by the 1290s. Some sections of Sand Canyon were abandoned and used as midden dumps, but the final abandonment occurred in a hurry. Large numbers of clay pots and stone tools still lie where they were left by their owners. Some kivas contain traces of abandonment rituals, as if their owners made a deliberate decision to move elsewhere. Some of the departing Anasazi left large, hard-to-carry utilitarian objects behind because they were going on a long journey. By 1300, the great Anasazi pueblos of the Four Corners were silent. The people had dispersed widely and joined distant communities elsewhere.

For two hundred years the Colorado Plateau saw tremendous cultural upheaval. The Anasazi largely abandoned the San Juan drainage as they dispersed into less affected regions. Many households moved into the Little Colorado River drainage, the Mogollon Highlands, and the Rio Grande Valley. Their descendants developed sophisticated irrigation agriculture, refined their dry farming, and dwelt in a great variety of settlements of all kinds. They developed new social and religious ideas over many generations of cultural uncertainty, until stability returned with improved environmental conditions after A.D. 1450.

The Anasazi themselves may have vanished from the cultural map, but their institutions and brilliant achievements are part of the fabric of Southwest society today. They survived the ravages of drought and El Niño episodes thanks to an ideology that placed movement and flexibility at the center of human existence. In the words of a Tewa elder: "They started coming and moving and then they settled and they stood up again and then they started moving again."[3] Still, we have no idea what price in starvation and suffering they paid for generations of living close to the environmental limit. After centuries of plenty and short generational memory of the effects of drought, the Anasazi were trapped in their towns and could only move away to survive. They at least had that option, and the ties of kin to support them in doing so.

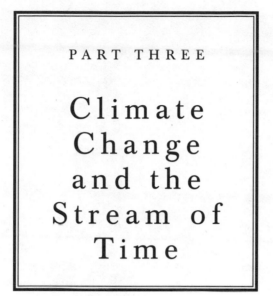

PART THREE

Climate Change and the Stream of Time

I don't, in all logic, see how any society can improve its lot when population growth regularly exceeds economic growth. . . . We will not slow the birth rate until we address poverty. And we will not protect the environment until we address the issue of poverty and population growth in the same breath.
  —The Prince of Wales, April 22, 1992

In the western world there are many material goals: economic growth, social welfare, better transport, more leisure and so on. But for our fulfillment as human beings we desperately need not just material challenges, but challenges of a moral or spiritual kind. . . . An appropriate challenge for everybody, from individuals, communities, industries and governments through to multinationals, especially for those in the relatively affluent Western world, is to take on board thoroughly this urgent task of the environmental stewardship of our Earth.
  —John Houghton, **Global Warming**

CHAPTER TEN

The Little
Ice Age

*I came to an immense mass of ice. . . . It was two or three pikes
thick, and as wide as the range of a strong bow. Its length stretched
indefinitely upwards, so you could not see its end. To anyone looking . . . it was a horrifying spectacle, its horror enhanced by one or
two blocks the size of a house which had detached themselves from
the main mass.*

—Sebastian Münster, on a Rhone glacier, August 4, 1546

Only 150 years ago, Europe came to the end of a 500 year cold snap
so severe that thousands of peasants starved. The Little Ice Age[1]
changed the course of European history. Dutch canals froze over for
months, shipping could not leave port, and glaciers in the Swiss Alps
overwhelmed mountain villages. Five hundred years of much colder
weather changed European agriculture, helped tip the balance of political power from the Mediterranean states to the north, and contributed to the social unrest that culminated in the French Revolution. The poor suffered most. They were least able to adjust to
changing circumstances and most susceptible to disease and increased mortality. These five centuries of periodic economic and social crisis in a much less densely populated Europe are a haunting re-

minder of the drastic consequences of even a modest cooling of global temperatures.

The Little Ice Age was the most recent of three relatively long cold snaps during the past ten thousand years. The Younger Dryas that triggered agriculture in southwestern Asia was the most severe, for it brought glacial conditions back to Europe. Another cold snap in about 6200 B.C. lasted four centuries and caused widespread drought. The Little Ice Age had more impact on history than its two predecessors, for it descended on the world after centuries of unusually warm temperatures. One can reasonably call it the mother of all history-changing events.

El Niños have destroyed civilizations and caused unimaginable suffering for at least five thousand years. They lie at one end of the climate-change spectrum—short, often severe events that roll across the tropical regions of the world and leave destruction in their path. By overthrowing powerful rulers and entire societies, such events have been as dramatic in their historical impact at the local level as the much longer climatic oscillations, measured in centuries, that have affected entire continents. All these fluctuations, whether El Niños or La Niñas, cycles of unusually stormy weather, or suddenly much colder temperatures, are part of a complex global climatic machine that includes oscillations on all scales. We know the machine is driven at least partially by complex interactions between the atmosphere and the oceans, and by deep-ocean circulations that transfer warm and cold water from the tropics to higher latitudes and back again. But many of the connections between such phenomena as El Niño and longer-term cycles such as the Little Ice Age remain a complete mystery.

This chapter changes climatic gears and tells the story of four centuries of unusually cold weather that altered the course of European history. The Little Ice Age operated on a different scale from a short-lived El Niño; it danced to a different climatic drummer than the protean Christmas Child. We do not know what caused this, or earlier, cold snaps, beyond a suspicion that deep-ocean circulations and arctic downwelling were important parts of the climatic equation. We know more about the causes of El Niños than we do about the much longer Little Ice Age.

The Little Ice Age was not a monolithic deep freeze, but a period of constant, and sometimes remarkable, climatic shifts between torrid summers and subzero winters. Like the Pacific, the Atlantic has a pressure oscillation of its own. A high North Atlantic Oscillation pressure gradient brings rain, strong westerly gales, and warmer temperatures. Cold years come when the North Atlantic Oscillation is low. These periods of warmer and colder conditions, which came and went during the Little Ice Age with unpredictable rapidity, were caused by complex and still little-studied phenomena such as movements of the Intertropical Convergence Zone at the equator and complex atmospheric interactions in the tropical Atlantic and Pacific. In another generation, many of these subtle teleconnections will be better understood. We will then discover whether the constant seesaw of the Southern Oscillation on the other side of the world played a role in shaping European history. In the meantime, the four centuries of the Little Ice Age offer an instructive lesson in the ability of long-term climatic oscillations to change the course of history in both gradual and sudden ways.

As Maya civilization collapsed in A.D. 900 and the Anasazi suffered through the great drought of the twelfth century, Europe enjoyed five and a half centuries of warmer temperatures and ample rainfall, commonly called the Medieval Warm Period. Average temperatures in the British Isles between 1140 and 1300 were up to 0.8 degrees Celsius higher than those of 1900 to 1950. Only today are some summer temperatures reaching Medieval Warm Period levels.

Greenland ice sheets tell us there was a burst of warmer weather in the far north between A.D. 600 and 650, followed by a more prolonged warm period that began about 800 and climaxed between 1150 and 1300. Norwegian farmers grew wheat north of Trondheim, at an unprecedented sixty-four degrees north. English vintners planted grapes as far north as Herefordshire in western England at an altitude of 200 meters above sea level. Landowners in the Lammer-

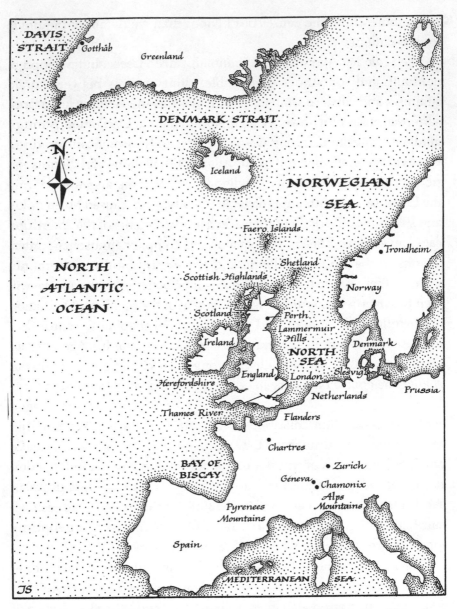

FIGURE 10.1   Map showing places mentioned in Chapter 10.

muir Hills of southeastern Scotland grew crops at 425 meters above sea level, during a golden age of Scottish history when interclan warfare was virtually unheard of. A burst of cathedral building spread across Medieval Europe in the twelfth century. Chartres Cathedral, built in a mere quarter-century after 1195, is a miracle in glass and stone, where ten thousand worshipers from the surrounding countryside once gathered on festival days to pour out their love for God. Chartres and its contemporaries were celebrations of the bounty of the soil, of generations of prosperity.

These were the climatically benign centuries when the Norse colonized Greenland and voyaged west to North America, William the Conqueror landed in Britain and imposed Norman rule, and Inca civilization rose to prominence in the high Andes. Warmer temperatures and higher sea levels affected New Zealand and the southwest Pacific, perhaps stimulating widespread Polynesian voyaging. It may be no coincidence that canoe travelers from the Society Islands settled New Zealand's North Island during this period, although the exact date of first settlement is unknown.

But the climate became more erratic during the thirteenth century. Alpine glaciers began to advance, and seasonal temperature changes became more extreme. As Arctic regions cooled, the thermal contrast between the Greenland-Iceland region and middle Atlantic latitudes steepened, causing greater storminess. Great westerly gales conspired with the prevailing high sea levels to cause vast destruction. Powerful wind storms and surging sea floods inundated low-lying North Sea coasts, drowning hundreds of thousands of people in some of the worst weather disasters ever recorded. The floods of 1240 and 1362 saw over sixty parishes in southern Denmark's diocese of Slesvig "swallowed by the salt sea." To add to the difficulties, tidal ranges increased after 1300, reaching a peak in 1400.

The Little Ice Age had begun.

The colder conditions of the Little Ice Age were not confined to Europe and North America. The world was on average one or two degrees Celsius cooler than it is today (during the late Ice Age it was six to nine degrees cooler). Precisely dated stalagmites from Cold Air Cave as far away as northern South Africa provide evidence of cooler temperatures during the Little Ice Age. Glaciers advanced, tree lines fell, and seas cooled.

The twelve-kilometer Franz Josef Glacier in New Zealand's Southern Alps once reminded me forcibly how much glaciers moved during minor climatic shifts like the Little Ice Age. I walked up to the glacier face, which thrusts down a deep valley backed by mountains that rise to the precipitous heights of Mount Tasman, 3,494 meters above sea level. The path wound through the barren valley floor and across fast-running streams fed by the melting glacier, native rain forest rising on either side. Changes in the vegetation on the valley walls marked the limits the ice had reached in recent centuries. We wended our way between massive blocks of hard rock polished into humps by the abrasive debris collected by the ice as it advanced and retreated along the valley. The slowly retreating glacier front glistened in the hot sun, a shadow of its former self even a century before. Franz Josef is a glacier on the move.

Eight centuries ago, Franz Josef was a mere pocket of ice on a frozen snowfield. Then the Little Ice Age began, and the glacier thrust into the valley below, smashing into the great forests that flourished there and coming within three kilometers of the Pacific. Today signs mark its retreat. Between about 1850 and 1893, the ice retreated rapidly as the Little Ice Age ended. The twentieth century has seen advances and retreats, with a cumulative fall back of about one kilometer from its front between 1871 and 1946. The occasional advances in this century, a few hundred meters over two or three years, are temporary interruptions in a steady retreat up the valley that will be reversed only by another oscillation in global climate.

The Little Ice Age was nowhere near as severe as the thousand-year-long Younger Dryas that triggered agriculture ten thousand years ago. At its height, between A.D. 1550 and 1700, mean temperatures were 1.2–1.4 degrees Celsius below those of the Medieval Warm Period.

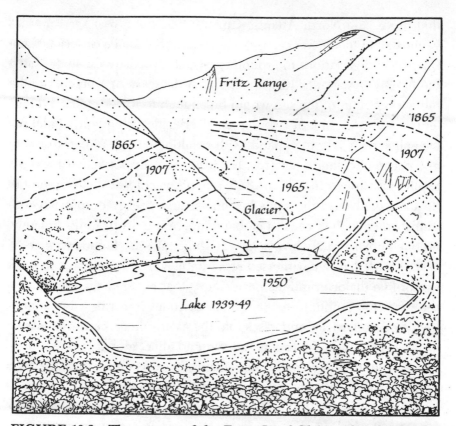

FIGURE 10.2   The retreat of the Franz Josef Glacier, South Island, New Zealand, 1865–1965. The glacier face has retreated even farther since the 1960s. Redrawn from New Zealand government sources; see also Jean M. Grove, *The Little Ice Age* (London: Routledge, 1988).

What caused the Little Ice Age? Scientists are still deeply divided over that question. All agree that it was an episode of profound importance to modern understanding of climatic change, for it unfolded during six hundred years of recent history when humanly caused global warming was not a potential factor.

As we have seen, the complex circulation patterns and interactions of the atmosphere and ocean are a major driving force in short-term climatic change. Could a massive influx of freshwater into northern waters from thawing glaciers and sea ice have turned off the "switch" that drives the North Atlantic Current and the Great Ocean Conveyor Belt in the deep ocean, as scientists like Wallace Broecker theorize? Or did variations in ground-level solar intensity cause the Little Ice Age? The scientific jury is still out. A decrease of 1 percent in the radiation output of the sun would be enough to lower global average temperatures by one or two degrees Celsius, causing continental sea ice and snow cover to expand, with resulting changes in atmosphere and ocean circulation patterns. For instance, sunspot activity was exceptionally low from 1715 to 1845, one of the coldest spells of the Little Ice Age. Could the sun's brightness and energy output have diminished enough to cause a global temperature drop? Unfortunately, no one knows exactly how solar variability affects climatic change.

The Little Ice Age witnessed remarkable volcanic activity: an average of five major eruptions per century that equaled the intensity of the Krakatoa eruption in 1883. Such episodes inject massive quantities of microparticles and gases into the atmosphere, causing massive dust veils that dim the moon and sun and affect global temperatures. The ash content of a central Greenland ice core shows that the years of the Medieval Warm Period between 1100 to 1250 were quiet volcanically. Between 1250 and 1500 and between 1550 and 1700, however, there were many eruptions, including a massive one in 1600 at an unknown location. Many scientists are sure that volcanic activity produced brief climatic extremes during the five cold centuries. They know the eruption of Mount Pinatubo in the Philippines in 1991 lowered the world's average temperature by about one degree for two years. They also point to 1816, "the year without a summer," as proof of the power of volcanic activity.

Between April and June 1815, Mount Tambora, a volcano on the island of Sumbawa in Indonesia, erupted massively. The explosion was heard in Sumatra, sixteen hundred kilometers away. Only twenty-six of the island's twelve thousand people survived. Ash clouds fell in Java, over five hundred kilometers away. Tambora was the largest volcanic eruption in modern times. The exploding volcano pumped into the atmosphere ten times the amount of ash produced by the notorious Krakatoa. Huge quantities of dust and sulphur dioxide produced a reverse greenhouse effect, forming in effect a sunscreen around the earth. Europe and North America shivered in 1816. Glaciers in the European Alps advanced vigorously. Snow fell in New England in June and July, crops failed throughout Europe, and famine was widespread. The raw summer weather caused the vacationing English poet Percy Bysshe Shelley and his wife Mary to stay indoors during their holiday on the shores of Lake Geneva. They entertained their friends by telling horror stories. Mary dazzled her audience with one called *Frankenstein,* about a monster that perishes with its creator in the frozen arctic.

Perhaps a dimmer sun and intense volcanic activity were players in the Little Ice Age equation. Whatever its cause, the five centuries of cooler weather brought profound changes in European history.

In about A.D. 982, the Norseman Eirik Thorvaldsson the Red sailed west from Iceland to explore the mysterious lands that sometimes appeared on the far horizon when the winds blew from the north. Three years later he and his men returned with stories of a fertile uninhabited land where fish were plentiful and the grazing grass lush and green. Eirik named it the Green Land, "for he said that people would be much more tempted to go there if it had an attractive name."[2]

Eirik persuaded twenty-five shiploads of settlers to sail for Greenland in 986. They founded the village of Brattahlid in the southwest and a "Western Settlement" at Gotthåb some 650 kilometers to the north. Life was never easy. The restless and adventurous Greenlanders occupied lowland areas around the inner shores of fjords, where they could pasture their stock and perhaps attempt to grow corn. All animals lived inside during the long winters, when the settlers lived off dried meat, fish, and stored dairy products. Their staple diet was seal meat, collected by the ton when harp seals migrated northward along the west coast in May and June.

The Greenland Norse always lived on the edge. Their lives depended on making full use of seasonal migrations of harp seal and caribou to obtain winter meat supplies. During the summers, the settlers mounted polar bear– and walrus-hunting expeditions to Nordsetur, the "northern coast" around Disko Bay, more than eight hundred kilometers north of the main western settlements. Bearskins and walrus tusks were the only trade goods of interest to the outside world. The Greenland Norse paid their annual church tithe to distant Norway in walrus tusks. Sometimes the tithe was more than four hundred tusks, far more than they could collect around their own settlements.

The colonists were expert seamen who explored every fjord and bay of western Greenland. Very early on, bold young men ventured far north toward the arctic ice and across the foggy and hazardous Davis Strait to the *Ubygdir,* the "unpeopled tracts," new lands beyond the western horizon.

During the 990s, Leif Eiriksson, son of Eirik the Red, sailed across to Baffinland, then southward in front of a northeast wind along the Labrador Coast to Newfoundland and the mouth of the Saint Lawrence River. Eiriksson wintered over in a wooded land he named Vinland, after the wild grapes that grew there, perhaps in Passamaquoddy Bay in northern Maine. The following year he and his thirty-five men returned safely to Greenland with a full load of timber.

The Norse never settled permanently in North America. They could survive on harsh Greenland coasts as long as the climate was relatively predictable, but they lacked the numbers and resources to expand and maintain pioneer settlements far to the west, where they had to compete with large indigenous populations and sail on ice-strewn, hazardous seas. Nor were there strong motives for colonization—such as religious persecution at home, or promises of gold and fertile land to attract greedy adventurers. So the Norse voyaged westward sporadically in search of timber, which was in short supply in Greenland.

The Greenlanders depended on the constancy of temperature swings and ice conditions from one season to the next. Even small perturbations greatly affected the abundance of food. During a cycle of slightly cooler years between 1954 and 1974, for instance, summer harp seal catches in the Gotthåb area of western Greenland declined sharply. Further north, the Kapisigdlit station, situated in a sheltered fjord, saw the percentage of harp seals in the annual catch fall from 30 percent to a mere 4 percent. If even this minor cooling caused such a profound drop, we can only imagine the consequences of a more prolonged and severe cold snap on a population living close to the edge. At the same time the harp seal catch was failing, longer winters would have required that domesticated animals be kept indoors longer in years when a shorter growing season had yielded much less hay. Colder and longer winters could also have deepened the snow cover, leading, in turn, to a dramatic reduction, even the temporary extinction, of caribou in parts of southwestern Greenland.

With their food base thus contracted, Norse fortunes declined rapidly. Malnutrition and premature deaths plagued even well-established settlements. Isolated communities became more vulnerable to attack from hostile Inuit groups. Meanwhile, the Inuit flourished in the cooler conditions, for they had adapted to Greenland's harsh and unpredictable environment for many thousands of years. They wore layered, tailored skin clothing that allowed them to hunt in subzero temperatures. Light Inuit skin boats and kayaks were ideal for operating in ice-strewn waters. Their harpoon technology, fashioned from bone and ivory, was among the most sophisticated in the world, so they could hunt cold-loving ring seal and fish through the ice in the depths of winter. Unlike the Norse, they were not solely dependent on the summer-migrating harp seal.

Unable to hunt sea mammals in winter, and apparently reluctant to change their lifeway, the Norse succumbed to climatic stress. By 1350 they had abandoned the Gotthåb settlement, perhaps after an attack by local Inuit. By 1500 the larger eastern settlement was also empty. When the ice spread farther south and endangered the most direct sailing route to the Green Land, even the most tenuous links evaporated as the bitter cold of the Little Ice Age caused major economic disruption throughout Europe.

Like the Greenlanders, Icelanders were at the mercy of sudden climatic changes. They subsisted mainly on fish and cattle, so land and sea temperatures and hay harvest yields were of vital concern. Hay grass is highly sensitive to air temperatures. Colder-than-average winters with intense frosts and deep snow cover can retard growth or even kill off the grass crop before summer. The soil may stay frozen until late spring, when a quick thaw floods the ground and kills the new grass all at once. In the exceptionally cold 1967 growing season, for example, hay production fell by one-fifth, with yields of 870 kilograms less per hectare than normal. Experts have calculated that a temperature deviation of one degree Celsius from the 1901–1930 norm of 3.2 degrees Celsius reduces the carrying capacity of the land by 30 percent. Despite modern farming practices, the Icelandic government still has to purchase and transport hay in cold years, at enormous expense.

Sea temperatures around Greenland and Iceland dropped precipitously for much of the time between 1600 and 1830, decimating cod populations, another staple of the Icelandic diet. Cod flourish in waters between two and thirteen degrees Celsius, but their kidneys do not function in colder water. Even a minor shift in polar water causes the fish to follow warmth. The Norse had subsisted off cod during the heyday of their settlements in Greenland, but there were no stocks off Greenland during the Little Ice Age. Cod disappeared completely from the Norwegian Sea during the seventeenth century as polar water spread southward.

Iceland has exported fish since the fourteenth century, although the size of cargoes was limited until the introduction of decked ships in 1890. But the industry has always been at the mercy of cooler sea temperatures. Even with modern industrial-scale fishing, herring and cod catches rise and fall with water temperatures. One of the major reasons for Iceland's bitter confrontations with Britain in the 1960s over fishing rights was the deterioration of fish stocks around Iceland as a result of falling sea temperatures. Iceland's dependence on cod and herring has made it vulnerable to sudden climatic change and resulted in firm, even extreme, political stands on fishing rights.

Until the onset of the Little Ice Age, the Icelanders also grew a hardy strain of barley in the north, south, and southeast of their homeland. However, the farmers had abandoned barley cultivation in the north by the end of the twelfth century. By the fifteenth century, no one grew cereal crops. Despite occasional experiments, barley did not return for eight centuries.

The Little Ice Age caused great suffering in Iceland from the seventeenth to the nineteenth century, a period during which mountain glaciers advanced, hay crops fell sharply, and thousands of cattle died of hunger and cold. In 1757 the sheriff of Salasysla in the northwest reported that "just in this year 21 cows and bulls, 1,292 sheep, 3,209 young lambs, and 151 horses have died in this one district. Forty-five people have died of hunger and wretchedness, and 15 dwellings have been deserted." He also reported poor fishing and noted that it "will be a pure miracle if a third of the population does not die of hunger."[3] The Icelanders fished from open boats. Even when they could launch their vessels into the ice-strewn water, they could not venture far from land. In the brutally cold years from 1750 to 1758, many fisherfolk moved inland and descended on hungry farming relatives. Nearly seventeen thousand people out of an island population of fifty thousand souls perished of hunger and associated ailments.

The start of the Little Ice Age had an immediate effect on European agriculture. Northern European vineyards went out of production between 1300 and 1310. Between 1313 and 1317, a series of exceptionally wet and unusually cold summers caused widespread crop failures and famine that killed thousands of people. There were outbreaks of cannibalism, and entire villages were abandoned or their populations decimated. The wet, cool summers and disastrous harvests undermined the viability of many small farming villages. Thirty years later the Black Death savaged Europe. Many of the hardest hit were those weakened by earlier famines.

By the beginning of the fifteenth century, most northern European farmers had abandoned wheat cultivation altogether. Wheat was a tricky crop at the best of times in the north, requiring constant tilling, frequent and careful manuring, and meticulous rotation from field to field. English visitors to a Danish wedding in 1406 commented on the widely sodden ground and lack of wheat fields. In Norway, where the cold weather and plague had reduced the population by two-thirds over the preceding century, upland farms lay deserted. By the 1430s many district tax yields had fallen to one-quarter of their level in 1300. Many poor families ate rye bread, which, a French doctor wrote in 1702, "is not as nourishing as wheat and loosens the bowels a little."[4] However, the bread crops never created abundance. Farmers ate beans and peas and made flour from buckwheat or chestnuts. Cattle were as important as cereal crops, for they provided meat and milk as well as manure. But the farmers were caught in a vicious circle, for they needed animals to draw plows and to fertilize the soil. Their beasts in turn required more grazing land at the expense of cultivated fields. A fourteenth-century almanac adjured the farmer to "multiply his livestock for it is this which will give the land the manure that produces rich harvests."[5] Nor did crops like barley, oats, and rye produce higher crop yields. The plants choked one another. From the fourteenth to the eighteenth century, Europe's peasants subsisted on coarse soups and gruels, as they had in prehistoric times.

The 1430s saw long spells of severe winter weather interspersed with very dry, hot summers and exceptionally wet springs and falls that caused havoc from England to the Alps. The Scottish Highlands erupted in warfare between hitherto peaceful clans. The severe winters reduced many Highlanders to making bread from tree bark. The murder of King James I, while hunting near Perth in 1436, was a direct result of famine-related social disorder, an event that caused the capital to be moved to the fortress at Edinburgh.

By 1500 European summers were about seven degrees Celsius cooler than they had been during the Medieval Warm Period. The growing season in England was shortened by about three weeks, and by as much as five by the seventeenth century. At the same time, the ground grew wetter. Marshes spread, and rivers flowed more strongly, making agriculture even more of a struggle. In northern

Norway, the highest altitude level at which farming was possible fell by at least 150 meters between 1300 and 1600. Fortunately, the warm Atlantic currents had kept the Icelandic and Norwegian fisheries operating when agriculture was in decline. During the early 1500s the climate warmed up somewhat, giving momentary relief. Wheat yields and land values increased slightly, only to fall again as the Little Ice Age reached its climax between 1550 and 1700. This time, temperatures in northern seas dropped sharply and cod catches fell dramatically, adding to the distress.

Some of the greatest suffering came in the shadow of the Alps. In June 1644, a procession of three hundred people led by the bishop of Geneva, Charles de Sales, made its way high in the Alps to "the place called Les Bois above the village where hangs, threatening it with total ruin, a great and terrible glacier come down from the top of the mountain."[6] The villagers had good reason to worry, for the Les Bois glacier was advancing "by over a musket shot [120 meters] every day, even in the month of August." The bishop duly blessed the glacier and repeated his invocations at a whole ring of ice sheets, which hemmed in seven small villages. It was as well that his blessings worked and the glaciers retreated, for the Les Bois glacier had blocked the valley of Chamonix itself and threatened to transform it into a lake. A quarter-century later, the ice sheets had retreated, but "the land they occupied [was] so barren and burned that neither grass nor anything else has grown there."

Still, the worthy bishop's efforts had little immediate effect. Between 1640 and 1650, a decade with cool and extremely wet summers, glaciers throughout the European Alps advanced farther than at any time since the Ice Age. In desperation, the people again prayed for mercy. By September 1653, the Aletsch glacier threatened so much farmland that the local people asked the Jesuits for assistance. Fathers Charpentier and Thomas preached reassurance to the community for a week. Then a solemn procession walked for four hours, bareheaded in the rain, to the "snake-shaped" glacier. The supplicants heard mass and a short sermon, before the priests sprin-

kled the front of the glacier with holy water in the name of Saint Ignatius and recited "the most important exorcisms." "On that very
spot just in front of the glacier, they set up a column bearing his ef-
figy: it looked like an image of Jupiter, ordering an armistice not just
to his routed troops, but to the hungry glacier itself."

We are told that Saint Ignatius stopped the glacier in its tracks, but
rapidly advancing Alpine ice sheets continued to threaten farming
communities in the foothills. In the eastern Alps the expanding Ver-
nagt glacier repeatedly dammed river valleys and formed lakes behind rubble barriers that broke repeatedly, flooding everything
downstream. At Christmas 1677, one village burned a vagrant suspected of practicing magic to block the valley. Hundreds of people
died in unexpected, catastrophic floods. Farming populations fell
sharply as people fled the encroaching ice. The hovering glaciers in
the Chamonix area brought bone-chilling cold, perennial frosts, and
"such strong winds that they sometimes carry away part of the hay
and grass after it has been cut."

The parish records of tiny mountain communities high in the Alps
chronicle great suffering during the Little Ice Age. The French historian Le Roy Ladurie has likened the fluctuations of Alpine glaciers to
the endless cycles of ocean tides. After centuries of "low water" during the Middle Ages, the ice sheets were high in the mountains.
Then, around A.D. 1300, the tide began to rise and the glaciers
spread downslope. A glacial "high tide" brought the ice deep into
foothill valleys between 1590 and 1850. The greatest thrusts occurred
in the seventeenth century and again in 1818–1820 and 1850–1855,
scarring villages and decimating Alpine pastures. By 1860 the tide
had turned and a great retreat began. By 1900 many glaciers had receded more than two kilometers deeper into the mountains in just
forty years. In the 1990s the tide is still out, as we live through some
of the warmest weather in six centuries.

A few contemporary weather records serve to show how extreme
conditions were. With the southward spread of arctic water into

northern seas, ice occasionally blocked the Denmark Strait between Iceland and Norway, and in 1695 it even mantled the entire Norwegian coast. Winter temperatures in England were up to one and a half degrees Celsius lower than the average for 1900–1950. In 1608 several meters of ice covered the Thames River. The exceptionally cold winter of 1684 froze the ground in parts of southern England to a depth of over a meter, while ice formed a five-kilometer belt along the English Channel coast. The length of winter frosts was much longer and the depths of snow cover much higher than in the twentieth century—up to 112 days at Zurich, Switzerland, in 1684. Interestingly, the most destructive weather often came in March, which in the 1880s was as much as three degrees Celsius colder than the twentieth-century average. This had serious consequences for local farmers, who ran out of hay and had to feed their cattle on pine branches and straw. They slaughtered many of their beasts while also enduring poor harvests, caused in part by a parasite that was active under snow cover and attacked growing corn.

Not that it was always exceptionally cold. There were some very mild winters during the height of the Little Ice Age, often soon after extremely cold years. Europe enjoyed tremendous variation during these centuries: The standard deviation of winter temperatures in England and the Netherlands was about 40 percent to 50 percent greater during the severest centuries of the Little Ice Age than during the early twentieth century, when westerly winds from the Atlantic Ocean dominated the weather pattern. There were occasional summer heat waves, too, notably in late June and July 1665, when the plague decimated London. The following year saw the twelfth-warmest summer temperatures in the previous 320 years—and the Great Fire of London.

Despite their hardships, Londoners made the most of cold winters. They skated and partied on frozen rivers and used them as roads. On January 24, 1684, the diarist John Evelyn wrote of "frost . . . more & more severe, the Thames before London was planted with bothes [booths] in formal streets, as in a Citty. . . . It was a severe judgement on the Land: the trees not onely splitting as if lightning-strock, but Men & Catell perishing in divers places, and the very seas so locked up with yce, that no vessells could stirr out, or come in." The still, cold air trapped ascending coal smoke, causing serious air pollution. The "fuliginous steame of the Sea Coal" prevented one from seeing across the street and filled Londoners' lungs with "grosse particles." Evelyn heard stories of extremely cold weather as far south as Spain "& the most southern tracts."[7]

As Londoners partied, the Scots suffered grievously. Up to one-third of the upland population died between 1693 and 1700, during the years immediately preceding the union with England in 1707. The harsh climate may have helped make union inevitable. A contemporary official wrote:

> During these disastrous years the crops were blighted by easterly "haars" or mists, by sunless drenching summers, by storms, and by early bitter frosts and deep snow in autumn. For seven years the calamitous weather continued, the corn [i.e., grain] barely ripening. . . . Even in the months of January and February, in some districts many of the starving people were still trying to reap the remains of their ruined crops of oats, blighted by the frosts, and perished from weakness, cold, and hunger.[8]

Throughout northern and western Europe, the coldest years of the 1690s, the years around 1740, and 1816–1819 were periods of constant famine. The harvest of 1693 was the worst in western Europe since the Middle Ages. France became a "great desolate hospital without provisions." Crop failures resulting from unusually heavy rainfall, part of worldwide climatic anomalies perhaps connected to the great ENSO episode in the late 1780s and early 1790s, contributed to the unrest that led to the French Revolution.

Unexpected salvation came from South America. Before the discovery of the Americas, European farmers depended primarily on cereals like barley, wheat, oats, and rye. Bread and various porridges made from these grasses, like Scottish oatmeal, formed the staple diet of millions of peasants. Except for a few roots eaten as side dishes—carrots, parsnips, and turnips—they relied almost exclusively on cereals. Crops like barley and wheat, which grow on long stalks high above the ground, are vulnerable to strong winds, hail, and excessive rainfall. Birds and insects eat the ripening grain. Cereal agriculture could be extremely productive in warmer, Mediterranean climates, but in northern Europe extreme weather and the resulting crop failures brought severe cereal shortages. The agriculture that had flourished during the Medieval Warm Period was ill suited to the Little Ice Age.

The solution to endemic famine came in the form of an ugly tuber first domesticated high in the Peruvian Andes.

The highland Andean Indians grew at least three hundred varieties of potatoes. Like maize and tobacco, potatoes were carried to Europe by Spanish conquistadors. Europeans spurned the new root crop at first. Folktales alleged that the misshapen potato caused leprosy. Some Russian Orthodox priests proclaimed potato eating a sin and named it the "devil's plant." Prussian servants threatened to change masters if fed potatoes. For two centuries the potato was little more than a curiosity, grown in some monastery gardens and by some gourmets as a novelty food. Cultural resistance to the potato was so strong that farmers would willingly endure repeated famine rather than change their diet.[9]

Once introduced, however, potatoes flourished in northern Europe, an environment as cool and damp as their Andean homeland. Ireland was the first country to grow the potato on a large scale. Legend has it that the tubers arrived aboard shipwrecks from the Spanish Armada in 1588, but they did not become a staple until the late seventeenth century. Irish farmers discovered that a piece of land that would feed one person on wheat could feed two on potatoes. The population of Ireland almost tripled as a result, from 3.2 million in 1754 to 8.2 million in 1845. Had the Irish followed the Native American practice of diversifying their potato strains, they might have avoided the disastrous effects of the potato blight that caused the dreadful famines of the mid-nineteenth century.

From Ireland, potato agriculture spread eastward across Britain and into the Low Countries. The great economist Adam Smith drew attention to the potato in his *Wealth of Nations,* published in 1776, and predicted it would feed large numbers of working people, making men stronger and women more beautiful. By the late eighteenth century, Frederick the Great of Prussia was forcing his subjects to grow potatoes or starve. Other monarchs did the same. Their advisers had learned that potatoes yielded more nutrition for less work per hectare than any grain crop, over a growing season of three to four months as opposed to almost double that for cereals. Potatoes grew in a wider variety of soils, needed less attention after planting, and, unlike oats and wheat, did not require lengthy grinding and processing. They could be stored for up to a year and made into bread or all manner of different dishes.

Once adopted, the new crop rapidly became a staple. Between 1693 and 1791, grain consumption in Flanders alone fell from 758 grams per person per day to 475 grams as potatoes replaced about 40 percent of cereal consumption. Nutritional diseases declined throughout Europe. By the 1830s northern Europe had become a major economic force, partly because the potato had reduced the famine cycles so typical of the Little Ice Age. In France, for example, there were 111 famines between 1371 and 1791, sixteen of them in the eighteenth century alone. The potato effectively eliminated this catastrophic cycle. The productivity and reliability of potato farming helped increase Europe's population and freed more workers for nonagricultural employment—such as manning the factories of the Industrial Revolution.

The cold centuries ended in the 1850s, as the Industrial Revolution was at its height. The world entered a new era of warmer temperatures and less extreme climatic swings, apparently triggered by entirely natural causes. (Some experts do wonder whether the higher levels of carbon dioxide released into the atmosphere by the growing forces of the Industrial Revolution contributed to the warm-up.) The warming has continued to this day, interrupted by occasional colder episodes. The three severe winters of 1939–1942 frustrated Adolf Hitler in France and Russia. Between 1940 and 1975, the world's climate cooled very slightly despite increased carbon dioxide levels, prompting talk of an imminent Ice Age. Since the 1970s the warming has continued. Climatologists report that 1997 was the warmest year in the twentieth century, with 1998 promising to be as warm if not warmer. How much of this warmth is due to the burning of fossil fuels and other human activities is a matter of debate. Perhaps another Little Ice Age is less likely now than it would have been had not the burning of fossil fuels increased so dramatically during the twentieth century. But we would be foolish to assume that another Little Ice Age is an impossibility.

Five centuries of cold caused subsistence crises in Europe. Today the same landscape produces large food surpluses on an industrial scale. But elsewhere people still starve. Now Africa suffers from hunger, caused by drought, social disorder, war, and massive cultural changes.

CHAPTER ELEVEN

"Drought Follows the Plow"

In spite of the scantiness of the vegetation great herds and flocks were seen and the scrub forest and the grass is burnt, great fires crossing the countryside. Overgrazing and hacking in the forest that is left, annual burning, and sand invasion suggest the question: How long before the desert supervenes?
—E. P. Stebbing, *"The Encroaching Sahara..."*

I remember watching a group of Bemba men and women in northern Zambia attack their cleared fields with iron-bladed hoes. They had started work early in the day when the sun was low, fanning out in rows over dry and hardened soil thick with weeds and coarse shrubs. *Thunk! Thunk! Thunk!* The hoes struck the obdurate ground again and again, often dozens of times at the same spot. Weeks of backbreaking, unceasing work at the hottest time of year sapped all the farmers' energy. No plows or wheels made the task easier, nor was there any guarantee that the rains would arrive on time. Each family needed about three hectares of cultivated and fallow land to feed itself. The amount of labor required to keep those gardens in production is enormous even in good rainfall years.

Every African subsistence farmer I have ever talked to is philo-sophical. He knows that at least once in his lifetime he will suffer from drought and malnutrition.

Yet Africans have often triumphed over environmental adversity. They created great kingdoms—Ghana, Mali, Zimbabwe—traded indi-rectly with the far corners of the earth, and fashioned great art and flamboyant cultures. But many have always lived on the edge. Re-gions like the southern fringes of the Sahara are so arid and unpre-dictable that it took remarkable ingenuity and environmental knowl-edge to survive there. Famine and disease stalked farmer and herder alike even when population densities were low. In the twentieth cen-tury, when colonial governments upset the delicate balance of cli-mate and humanity on a continent that had never supported vast numbers of people or huge urban civilizations, these social disasters have come more frequently. Europe suffered through repeated sub-sistence crises during the Little Ice Age. Africa is enduring far more serious food shortages in the late twentieth century, caused by a com-bination of drought, population growth, and human activity.

The Sahara and the Nile Valley occupy about half the entire African continent, an area larger than the United States, including Alaska. The Atlas Mountains of Morocco and the Ethiopian Highlands rise at opposite ends of the desert. In the east, Saharan sand laps the Nile Valley, an oasis for animals and humans since the beginning of the Ice Age. The desert's southern margins pass gradually into stunted grassland, then dry savanna. Each year, the southern frontiers ex-pand and contract as droughts come and go or occasional cycles of higher rainfall allow dry grass to grow a few kilometers farther north.

The Sahara is one of the hottest places on earth. The prevailing wind, a dry, descending northeasterly, raises the temperature above thirty-seven degrees Celsius on more days of the year than anywhere else in the world. The heat causes enormous evaporation loss from surface water and vegetation in an area where the mean rainfall is less than 380 millimeters a year. Vast tracts of the desert have sup-ported no animal or human life at all for nearly five thousand years.

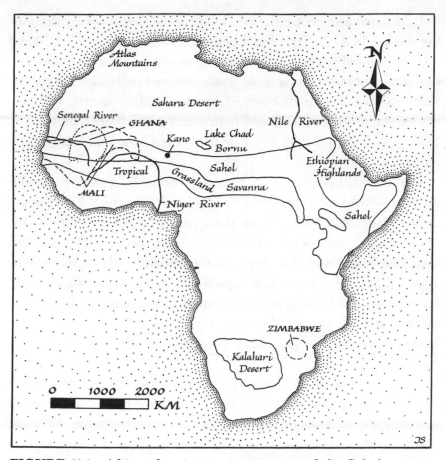

FIGURE 11.1   Africa, showing ancient states and the Sahel region.

The Sahara was not always so inhospitable. Long-term climatic changes during the Ice Age brought it higher rainfall. In warmer periods, enormous shallow lakes and semi-arid grassland covered thousands of square kilometers, nourished by seasonal rivers that radiated from the desert mountains. Lake Chad, on the borders of modern-day Chad, Niger, and Nigeria, is a climatological barometer of the ancient Sahara. Today it is minuscule compared with the Lake Chad of 120,000 years ago, which filled a vast basin larger than the Caspian Sea.

Lake Chad's ancient shorelines tell us the desert was very dry during the height of the last glacial period, from about 18,000 to 10,500 B.C. The global warm-up after the end of the Ice Age brought more humid conditions. During these better-watered millennia, the Sahara supported sparse populations of hunters and gatherers, who anchored themselves to lakes and permanent water holes. Not that the Sahara was a paradise. The climate was at best semi-arid, with irregular rainfall.

Around 5000 B.C., some of these tiny nomadic groups either domesticated wild cattle and goats or acquired tame beasts from people living in the Nile Valley or on the desert's northern frontiers. We know cattle herders once lived in the desert because they painted their beasts on cave walls high on the Sahara's central massifs. In 4000 B.C., at least three million square kilometers of the Sahara could have supported herders and, theoretically, as many as twenty-one million head of cattle. Of course, nowhere near that many cattle grazed there, but the pastoralist population of the arid lands grew steadily.

Sometime before 2500 B.C., for reasons that are still not understood, the Sahara began to dry up. A small drop in annual rainfall was sufficient to make the already arid land uninhabitable. As it had done in earlier millennia, the desert acted like a giant pump. During wetter times, it sucked people into its fastnesses to settle by shallow lakes and cool mountain ranges. The same groups migrated out to the desert margins when drier conditions evaporated the lakes and destroyed the grassland. As Ancient Egypt's first pharaohs strove to unify their kingdom, the Sahara's herders moved to the edge of the Nile Valley and southward into the Sahel, a savanna belt that extends across the continent from West Africa to the Nile River. Their remote descendants still graze their herds there to this day.

The word *Sahel* comes from the Arabic *sahil*, which means "shore," an apt description for a frontier land that borders the largest sea of sand in the world. The Sahel is a gently undulating grassy steppe between two hundred and four hundred kilometers wide, constrained to the north by the Sahara and to the south by a variety of forested terrain. The region just south of the Sahel is also the home of the tsetse fly, fatal to cattle and hazardous to people. Stunted grassland dominates at the margins of the desert, where the rainy season lasts a mere three to five months a year. Farther south, the grass gradually gives way to tree-covered steppe and eventually to more densely vegetated terrain. This inhospitable country, with large areas of sand dunes and long-dried-up lakes, supported cattle herders when Old Kingdom pharaohs were building the pyramids and when Julius Caesar conquered Gaul. Over many centuries, the herders developed effective strategies for surviving a highly unpredictable environment.

Most of the time the Sahel lives under a relatively arid weather regime. The subtropical high-pressure zone over northwest Africa brings dry conditions, while the north-south movements of the Intertropical Convergence Zone (ITCZ) cause rainfall. Globally, the ITCZ marks the convergence of the northeast and southeast trade winds. In the Sahel region, it forms the transition between the dry, northeasterly harmattan winds that blow across the Sahara and the moist southwesterly monsoon flow that originates in the South Atlantic. This convergence, with its heavy cloud cover and intense rainfall, moves northward during the Northern Hemisphere summer, bringing a short rainy season to the equator side of the Sahara. The rainfall at any given latitude depends on the number of months that the ITCZ dominates the local weather. The farther south from the desert, the longer the rainy season. Later in the year the ITCZ shifts southward again. High pressure fills in over the Sahara, and the dry season begins. These movements are part of atmospheric tropical circulation, the Hadley Circulation: Air rises near the equator and falls in subtropical latitudes, at about thirty degrees from the equator.

The scenario is, of course, more complicated than this, for the ITCZ is discontinuous in time and space. Weather disturbances move westward across Africa about every three to five days, either cloud clusters of intense rainfall or rapidly moving squalls that can move at

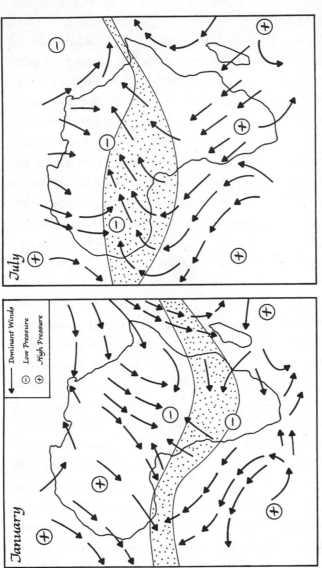

FIGURE 11.2 Movement of the ITCZ over Africa causes rainfall over the Sahara. In winter (left), the ITCZ drifts low across Zaire and East Africa. The Sahara receives dry, northeasterly winds. In summer (right), the ITCZ moves northward into the Sahel and southern Saharan region. The ITCZ separates the dry, northeasterly winds that blow across the Sahara and the moist southwesterly monsoon from the South Atlantic. During the northern summer, the ITCZ, with its heavy cloud cover and intense rainfall, moves northward, bringing a short rainy season to the southern margins of the Sahara. During the northern winter, the ITCZ shifts southward and no rain falls in the Sahel.

up to fifty kilometers an hour. This is weather wrought on a large scale, for the easterly waves originate over East Africa and can propagate west as far as the Atlantic or Caribbean, where they may generate hurricanes.

Rainfall in the Sahel is very erratic from year to year, though no one understands why. For years people assumed that the movements of the ITCZ were responsible. If the ITCZ did not move far enough north, went the argument, the southern margins of the Sahara remained dry. This does not explain, however, why rainfall diminishes not only over the Sahel but over Africa as a whole, especially the Kalahari and southern Africa. Scant rainfall in the Sahel almost invariably coincides with a similar shortfall at the other end of Africa, on the fringes of the Sahara's southern equivalent. The explanations most likely lie in much larger, global teleconnections, complex and still little-understood interactions between the atmosphere and ocean. As we have seen, drought conditions in southern Africa often coincide with strong ENSO episodes.

The El Niño years of 1972 and 1982–1983 also brought droughts to the Sahel. Far to the south, in Zimbabwe, acute droughts caused by El Niños in 1983 and 1992 caused severe crop shortfalls. The 1992 drought threatened as many as eighty million people in southern Africa with the early stages of famine. The dry spell hit after poor rains and below-average crops the year before had left many farming communities already hungry. Regional grain stocks in Zimbabwe were dangerously low because the International Monetary Fund (IMF) had fostered a "structural adjustment program" to reduce the country's budget deficit and rampant inflation while combating stagnant economic growth. In response, Zimbabwe exported much of its food reserves both to reduce government spending on food storage and to earn precious foreign exchange. Before the structural adjustment program, the country had six months' food reserves on hand, more than enough to deploy to famine areas. Now the government had to buy maize on the international market for prices as much as three times higher than earlier in the year. These purchases, and vigorous efforts by private agencies, averted widespread famine, but at a high price.

Like the Sahara, the Sahel was not always so arid. Fossil pollens show that the shores of Lake Chad supported a well-watered tree savanna around 3000 B.C. At the time, the Sahel may have averaged 650 millimeters of rainfall a year, as opposed to today's 350 millimeters. As increasing aridity pushed the Saharan herders to the margins of the desert, they settled at first in much better watered lands, where some herders already flourished. Then still drier conditions set in, lowering Lake Chad to near-modern levels, which have persisted with short-term fluctuations since then.

For the past twenty-five hundred years, the Sahel has enjoyed a climate characterized by irregular and sometimes severe droughts, as well as sharp variations in rainfall from year to year. Most precipitation falls in short, violent storms distributed erratically over a chronically dry region. The Niger and Senegal Rivers, and also Lake Chad, provide permanent water sources for those living in their vicinity, but cattle herders have to rely mainly on water holes, perennial streams, and shallow wells dug in riverbeds. It is not a life conducive to either rapid population growth or a high population density.

The chronicles of medieval Arab geographers and pollen data from Lake Chad hint that the Sahel was somewhat wetter between the ninth and thirteenth centuries A.D., during Europe's Medieval Warm Period. Arab travelers, and later Europeans, wrote of standing lakes and rich pastures in these centuries and after, but there is evidence of constant rainfall changes. For example, Lake Chad's water level fell by some five meters in the few decades before A.D. 1450, then rose by three meters by 1500, to fall again half a century later. The seventeenth century saw a level five meters above today's. Subsequently, the lake fluctuated between the modern level and shorelines as much as three meters over, with two major dips in the eighteenth century, one about 1850, another after 1913, and one more during the 1970s. Severe short-term droughts have arrived three or four times a century for the past few hundred years. The twentieth century is right on schedule. Without question, this chronicle of wetter and drier cycles, of longer- and shorter-term droughts and more plentiful rains, extends deep into the past.

The word *drought* assumes a complex meaning in the Sahel, where there is no such thing as an "average" rainfall, and where rain is highly localized, brief, and often violent. In any given year, as many as half of all modern-day reporting stations experience below-average rainfall. Almost no rain may fall throughout the wet months, or it may fall plentifully for some weeks but then fail completely, leaving young crops and fresh grazing grass to die in the ground. Furthermore, since most rainfall arrives as heavy showers during the hot months, losses from evaporation are enormous, little moisture sinks into the ground, and the potential for plant growth is much reduced.

Drought in the Sahel has both human and environmental dimensions. A single dry year may kill off stock, reduce grazing land, and devastate crops, but improved rainfall the next year will mitigate the impact. However, a succession of arid years may have a cumulative effect on cattle and humans, to the point that an unusually severe drought can deliver a knockout blow to already weakened communities. Sahelian dry cycles can persist for up to fifteen years, as can periods of higher rainfall. The latter lulls everyone into a false sense of security. Cattle herds grow, fields are planted ever farther north into normally arid land, contributing to the disaster if a long dry period arrives without warning. If anything is "normal" in the frontier lands, it is the certainty that severe drought returns. The ancient Sahelian cattle herders planned their lives accordingly.

Before the European powers carved up Africa in the late nineteenth century, the rhythm of cattle herding and agriculture on the Sahel changed little from year to year. The casual observer of a herding community might assume that the constant movements of people and herds were almost haphazard. In fact, the elders discussed every move carefully and acted conservatively. They relied on a remarkably detailed knowledge of the surrounding region, its water supplies and grazing grass, and were constantly in search of information about conditions elsewhere. The people were accustomed to inadequate diets and months of hunger when the rains failed and grazing grass or crops withered in the blistering sun. Over many generations, they had developed effective strategies for minimizing stock losses and surviving arid years. They depended on close family ties and enduring obligations of reciprocity with neighbors near and far to move their beasts, obtain loans of grain, and keep some of their animals safe from rinderpest and other catastrophic diseases.[1]

Groups like the Fulani used to move over large distances, sometimes two hundred kilometers or more. These movements coincided with the seasons. During the wet months from June to October, the herders would move north with the rains, feeding their cattle off fresh grass. As long as the grass ahead looked green, the herds migrated north. In a good year the cattle gained weight and gave ample milk. Nevertheless, the people moved constantly to find the best grass and to keep the cattle away from soggy ground. When the herds reached the northern limits of the rainfall belt, they turned south and consumed the grass crop that had grown up behind them as long as standing water remained.

Back in their dry-season range, the herds found grazing that could tide them over for the next eight months or so. Between November and March, many herds also grazed on farmland, rotating from one village to another, for their manure helped fertilize the soil for the next year's crops and their owners received millet in exchange. By the middle of the dry season, the herders were searching constantly for good grazing range, moving frequently and staying close to river floodplains if they could. The end of the dry season in April and May was the worst time of year. People were lethargic, often hungry, and deeply concerned about the condition of their beasts, who, even in a good rainfall year, might not have enough to eat. Then the first rains fell, green shoots carpeted the landscape, and the entire cycle began anew.

The herders' lifeway remained viable for thousands of years because they used effective strategies for coping with drought. They made best use of seasonal pastures and took great care in planning their moves. They moved their herds constantly to avoid overgrazing valuable range and spreading diseases like rinderpest. At the same time, they saw cattle as wealth on the hoof, so a herder sought to own as many head as possible, both to fulfill marriage gifts and other social obligations and to have as a form of insurance. The more animals he had, the more would survive the next drought.

While each nuclear family owned its own herds, they shared information over a wide area. They minimized risk of disease and drought by farming out the grazing of many of their animals to extended family and kinspeople living a considerable distance away. Like the !Kung, the Fulani and other herders placed a high premium on reciprocal obligations as a way of coping with unpredictable climatic conditions. They suffered but survived.

The twentieth century and the European powers changed everything. The French, who assumed control of most of central Africa from the local inhabitants after 1889, stressed the need for each of their new colonies to become financially self-sufficient, which meant taxing the people to pay for the cost of their government. At the same time, French authorities insisted on the free use of pastoral land by all and abolished the traditional political framework that had maintained control over grazing rights. This draconian policy opened the way for overuse of the best grazing ranges on the edge of the desert. At the same time, new colonial frontiers cut across traditional grazing ranges, further disrupting traditional culture. Sahelian population increased threefold thanks to well digging, improved medical care, and sanitation. National governments restricted mobility, initiated cash-crop planting schemes, and encouraged cattle herders to settle in permanent villages. The colonial economy had arrived on the Sahel. Granaries might bulge with grain, but the stockpile was sold for cash to pay taxes, making it impossible for many communities to feed themselves in lean years. When drought struck, the people had three options: suffer quietly and starve, find alternative food supplies, or move away to find food or earn cash to feed themselves. There was food available, but prices often skyrocketed as much as thirtyfold. Half to two-thirds of the herders were soon reduced to complete penury as they disposed of their animals and possessions under unfavorable terms. One thing never changed: the inexorable demand for scheduled tax payments.

In 1898 a prolonged dry cycle began, culminating in severe droughts in 1911 and 1914–1915. Lake Chad shrank by half, Nile flood levels fell by 35 percent, and thousands of cattle herders died of starvation. The famine remains seared into the memories of the Fulani, Hausa, and others.

Hunger gripped the Sahel with unrelenting severity. In 1914 grain reserves were low, livestock were already weak, and malnutrition was widespread, especially among children. By the time the growing season arrived, many people were too weak to work in the fields. Entire families died at the side of paths as they tried to flee south into better-watered areas. Those who survived did not have the strength to bury the dead. Many mothers abandoned their children in marketplaces, hoping some benefactor would feed them. Hundreds of younger people moved elsewhere to work on groundnut plantations in Gambia and Senegal. Families drove their cattle southward to sell them for gold, establishing a pattern of long-distance trade that persists to this day.

The 1914 famine was the first major challenge faced by British and French colonial administrators in their new Sahelian colonies. A handful of civil servants collected taxes and preserved imperial rule over an enormous area with only a handful of employees and minimal budgets. The Sahel was the obscurest of obscure postings at a time when much of the world was governed by almost forgotten commissioners, governors, and residents far removed from London and Paris. A famine in this part of the world did not cause a ripple of concern in a Europe preoccupied with a bloody world war. When the famine came, a colonial officer could do little except sit and observe the suffering around him. The British resident in Kano, Nigeria, reflected a widespread attitude: "Yes, the mortality was considerable but I hope not so great as the natives allege—we had no remedy at the time and therefore as little was said about it as possible."[2] A scattering of government notebooks, tax records, and tour reports tell a tragic tale. In Nigeria's western Bornu region, ten thousand Fulani had eighty-eight thousand head of cattle in 1913. A year later only fifty-five hundred people remained with a mere twenty-six thousand head. These figures compare well with the drought of 1968, when cattle losses exceeded 80 percent in the lower rainfall areas closer to the desert and between 30 and 60 percent in slightly better watered areas farther south. One administrator's report from Niger in 1913 spoke of "an important exodus to more favorable regions. A great number of cattle have died. . . . Animal mortality: cattle 1/3, sheep and goats 1/2, and camels negligible." An official collecting taxes in central Niger reported: "I have walked through 23 villages during this tour. . . . Last year these villages had a population of 13,495 taxable inhabitants. 3,354 persons have died during the famine."[3] Tax records reveal a 44 percent population decrease in one area of the country between 1910 and 1914. Another region reported a loss of 44,235 tax-paying adults out of a total of 57,626 in August and September 1914. Eventually, at least 80,000 people died in central Niger alone.

The cycles of wetter and drier conditions persisted. Another, less intense drought came in the 1940s. The 1950s and early 1960s saw greater rainfall. By this time every West African country in the Sahel had experienced sharp population growth caused by improved medical care, by a long-term campaign of digging artesian wells, and, above all, by the suppression of debilitating cattle diseases like rinderpest. Fifteen years of good rainfall led naive government planners to rash undertakings. They offered incentives to subsistence farmers to move into drier areas. Many of them were cattle herders

who had become cultivators. Dozens of communities moved northward into marginal lands where grazing grass had now sprouted.
Carefully rotated livestock could flourish here in good rainfall years,
but sustained agriculture without large-scale irrigation was another
matter.

As farmers moved in, the nomadic cattle herders shifted even
farther north, right to the frontiers of the desert. The nomads were
trapped between the Sahara and a rapidly growing farming population, which, in turn, was stripping the natural vegetation off land
that could support farming only for short periods of better rainfall.
Governments and international agencies both financed well-
digging programs, which provided ample water at the local level
year-round for large cattle herds. Between 1960 and 1971, the Sahel's cattle population rose to between eighteen million and
twenty-five million. According to the World Bank, the optimum is
fifteen million.

Then, with the inevitability of a Greek tragedy, the rains failed.
Drought first took hold in the north near the desert margins, then
moved southward in subsequent years. By 1968 rainfall in the more
marginal areas was less than half that of the 1950s as the drought extended well beyond the Sahel. The dry cycle persisted well into the
1980s, making the twentieth century one of the driest in the past one
thousand years.

This time nobody died of thirst, for they could drink from the deep
wells. But these same water supplies did not nourish the sparse natural vegetation. The Swedish International Development Agency,
studying cattle mortality in the early 1970s, found that most beasts
died not of thirst but of hunger, because their forage was gone. Thousands of cows clustered around the wells, then staggered away from
the water with bloated stomachs. They would struggle to free themselves from the clinging mud and often keel over from exhaustion.
Each well became the center of a little desert of its own, stripped of its
vegetation by thousands of cattle converging on the water source
from hundreds of kilometers around.

The herders, with their simple technology, had always maintained their herds in a finely tuned balance between natural water supplies and drought-resistant vegetation. Now the balance was gone.

The herders had too many cattle for the land. Overgrazing stripped away the deep-rooted, two-meter-tall perennial grasses that had fed the herds for centuries. Plants with shallower roots replaced them, then gave way to coarse annual grasses. When heavy grazing removed these in turn, next came leguminous plants that dried quickly. Finally, the herds pulverized the bare soil into fine particles that were blown by desert winds to the foot of slopes, where they dried into a hard cement. As the cattle ate the landscape, their owners felled every tree around for firewood. If this was not enough, encroaching farmers set fires to clear the land and killed off at least half the range grass each year.

Within a few years, grazing lands had become desert. At the same time, a rising farming population in the south had taken up just about all the cultivable land. Earlier generations, with plenty of fields to go around, could let much of it lie fallow and regain its strength. By 1968 there was no land to spare, nor did the villagers have fertilizers. Thousands of hectares of farmland yielded ever sparser crops until their owners just walked away, leaving a dustbowl behind them. Thousands of farmers and herders had no option but to flee southward. They descended on villages and cities, many of them ending up in crowded refugee camps.

At the time, many experts believed the expansion of the desert was an inevitable result of natural climatic change. In fact, human activity had degraded the marginal lands in ways unknown in earlier times. Food production declined throughout the Sahel, except on irrigated lands, where governments grew cash crops like groundnuts and grain for export. The drive for export crops pushed groundnuts into areas where drought-resistant millet was once a staple for village farmers. The millet farmers moved onto marginal land, so even more people went hungry. Throughout the crisis, West African governments still exported food while tens of thousands of their citizens subsisted on meager diets from humanitarian aid programs.

This particular drought cycle was unusually persistent. The climatologist Jule Charney believes the drought was reinforced by what he calls "biogeographical feedback": changes in the Sahelian land surface caused by the human impact on the fragile ecosystem of the arid lands. Interestingly, the same drought did not persist in the Sahel's southern equivalent, the Kalahari, where there are no dense herder populations. Numerical models tend to support Charney's hypothesis. They show that changes in the Sahel's land surface could indeed prolong drought. These changes could include a reduction of surface temperature caused by the exposure of soils with a high albedo through stripping of vegetation, by a reduction in surface moisture, and by higher dust levels over the desert, all of which tend to reinforce arid conditions. All these circumstances can result when overuse combines with extreme dryness, as happened in the 1968–1972 drought.

When French newspapers ran stories on the Sahel drought in 1972, the world took notice. Readers saw graphic pictures of emaciated children suffering from malnutrition and arid landscapes stripped by overgrazing and ravaged by years of inadequate rainfall. The governments of Chad, Mali, Niger, and other Sahelian countries were embarrassed by their inability to feed their own populations. Still, foreign governments and United Nations food agencies adhered to a long-standing policy of not interfering in the domestic affairs of member nations. Private famine relief agencies did not have the resources to provide assistance on the scale needed. When food and other supplies did arrive, it was too late for many people. Between 1972 and 1974, 600,000 tons of grain came to the Sahel from the United States alone, but the relief came late, was often unpalatable, and was distributed with great difficulty. The famine cost to foreign governments and private relief agencies was near $100 million. Between 100,000 and 200,000 people perished along with about 1.2 million cattle.

Thousands of hectares of marginal land suffered such severe soil erosion from overuse that they became useless for either farmers or herders. The drought could have been predicted, for centuries of experience showed that periods of ample rainfall were followed invariably by dry cycles. Nor was this drought particularly severe by historical standards. But thousands more people were affected.

What had changed was the much larger number of farmers and herders exploiting an already high-risk environment. The imbalance in the equation was the human one. The people in the affected areas lived by a fundamental tenet of the cattle herder. Wealth is cattle, and the more cattle you own the better. At worst, you will lose half your cattle in a drought, but more will survive than if you had not expanded your herds. Said one Fulani elder after the 1968 cataclysm: "Next time I will have two hundred [head]." He believed that one hundred would survive the next drought. However, the carrying capacity of the land is such that he would be lucky to have fifty survivors.

After 1973 the Sahel's farmers enjoyed somewhat better rains and obtained better crop yields. But drought conditions soon returned and endured until the mid-1980s. The 1982–1983 dry cycle, also coming in an El Niño year, was as intense as that of 1972. But international relief agencies had established themselves in the region during the earlier crisis. Severe food shortages developed across the Sahel, but famine gripped only parts of Chad, the Sudan, and Ethiopia, countries engaged in civil wars at the time. As always, war and social disorder aggravated hunger and humanitarian efforts. In 1998 the Dinka and other southern Sudanese people are suffering a major famine only because warring factions prevent adequate rations from reaching the needy.

Are the severe Sahel droughts of the late twentieth century aggravated by human activities? Do they result in part from overexploitation of the land, and in perhaps even larger part from humanly induced global warming? Although everyone agrees that higher Sahel populations are part of the problem, they differ widely on the effects of long- and short-term climatic change. One argument says that as the world gets warmer, arid areas like the Sahel and Kalahari will become even drier, while wet regions become even wetter, than they are today. Another school of thought hypothesizes that climatic zones will shift to make both wet and dry areas more arid. Rising populations and widespread environmental degradation make it hard for climatologists to develop reliable drought and rainfall forecasts.

The West African catastrophe was not caused by unusually severe drought or global warming. It resulted from uncontrolled population growth, careless and naive development planning that took no account of the lessons of history, and people making decisions about their environment in good faith for short-term advantage, without giving thought to the future. In the words of Randall Baker: "When the rains come in the Sahel, and the millet grows again, then the 'problem' will be considered over until next time."[4]

Western-style political and economic institutions have failed dismally in the Sahel. Over the past century they have brought repeated crises and famines, marginalized millions of people, and killed thousands. This failure points up the great achievement of traditional Sahelians in maintaining stable herder societies where modern economies cannot. For thousands of years the herders adapted successfully to their unpredictable and harsh environment. They maintained a detailed knowledge of grazing and water supplies over enormous areas, moved their herds constantly, and adjusted month by month to changing conditions. Their mobility, low population densities, and careful judgments gave them the ability to endure drought, the ravages of cattle disease, and constant uncertainties. The twentieth century and colonial rule brought longer life expectancies and better medical care, but rapid population growth and much larger herd counts followed in their train. The new regime also brought an economic system that fostered cash crops and imposed taxation to support cities and central government. When rainfall was slightly higher and cotton and groundnuts could be grown to feed the cash economy, the farmers and herders lost control of their own production and of their lives. Meanwhile, the same system pushed the herders and their growing herds into marginal areas, abolished their traditional grazing ranges, and made them dependent on the outside world for water and food. As Michael Glantz has aptly noted, "Drought follows the plow."[5]

The Sahel tragedy continues to unfold with dreadful predictability, a replay, on a different scale and in a different place, of the Maya collapse and the Anasazi dispersals. But this time there is nowhere to go.

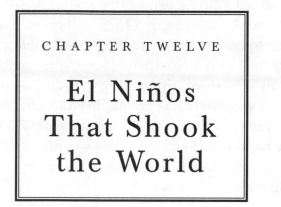

CHAPTER TWELVE

El Niños
That Shook
the World

It looks like an El Niño, it quacks like an El Niño. . . .
—Heard at a conference on El Niño, May 1997

It was 1983 when El Niño became an international media event, largely because Californians suffered through a winter of capricious storms and mudslides. In October few of us had heard of El Niño. Six months later we had lived through the most thoroughly monitored and most expensive El Niño in history.

The royal yacht *Britannia* lay in Long Beach Harbor, about to sail for Santa Barbara to bring Queen Elizabeth II to visit President Reagan at his nearby ranch. Sheets of rain poured from a lowering gray sky blown ragged by a thirty-five-knot southeast wind. Huge waves battered the Santa Barbara harbor breakwater and flooded roads. *Britannia* had planned to anchor close offshore, but conditions were suicidal off the town. There was talk of canceling the visit, but Her Majesty insisted on flying to her rendezvous. She traveled to the airport in a commandeered school bus and up the mountain to the Reagan ranch in a four-wheel-drive vehicle. "Just like a day on Dartmoor," remarked a British journalist, while his American colleagues talked of the "storm of a century." None of them mentioned El Niño.

The same storm generated a swirling tornado that stripped part of the roof off the Los Angeles Convention Center and ran up Broadway close to downtown, overturning cars and smashing in brick storefronts. Irene Willis was driving her Lincoln Versailles along Broadway when the tornado lifted the automobile into the air and dropped it hard onto the ground. The windshield blew out, and a piece of plywood lodged in the backseat. Live power lines fell on the car as Willis crawled to safety.

After the Gulf War, backward-looking scientists called it "the Mother of El Niños," a name worthy of the event. The exceptionally intense El Niño of 1982–1983 ravaged tropical forests in Borneo and brought floods to southern China and drought to Australia. Peruvian fish meal was so scarce that American farmers planted more soy as chicken feed. A poor harvest in the parched Philippines raised prices for coconut oil, soaps, and detergents. Renewed drought brought famine to the Sahel, and thousands of southern African farmers starved. Meanwhile, Californians watched expensive houses slide down waterlogged hillsides. In Peru the small coastal fishing villages of a century ago are now cities. The 1982 ENSO debilitated the anchovy fishery and sent roaring floods cascading through poor urban neighborhoods; hundreds drowned. The two great El Niños of the late twentieth century brought home a harsh reality: With millions more people on earth than even a generation ago, packed into cities or cultivating semi-arid lands and the margins of tropical rain forests, we are highly vulnerable to protean climatic shifts. It helps that science has made giant strides in forecasting global weather.

Climate models, computers, recording instruments, and satellites–the tools used to predict ENSO episodes–are products of a new multidisciplinary science that came into being during the 1970s. In 1972 major

225

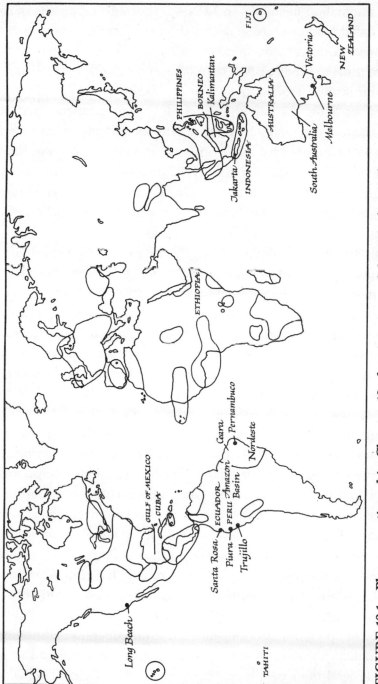

FIGURE 12.1 Places mentioned in Chapter 12, showing areas of drought (circled).

1972 major

climatic anomalies, including an El Niño, affected many parts of the world. Global food production declined for the first time in more than twenty years. Governments began to ask whether scientists could develop the tools to predict ENSO episodes and provide early warning systems for farmers around the world. The Sahel famine raised haunting questions about the human impact on the environment. Had humans expanded the Sahara by grazing cattle in a fragile environment, stripping the natural vegetation from semi-arid lands? Or were short- or long-term climatic shifts to blame? "Climate-related impact assessment" became a popular buzzword. This new science is revolutionizing our ability to forecast El Niños and other climatic anomalies, even if we cannot predict all their maverick swings.

During the 1970s a slight cooling trend had many experts proclaiming that the world would eventually plunge into another glacial period, in as few as twenty-three thousand years. They based their arguments on cyclical changes in the angle of the earth's rotation and on new evidence for past glacial episodes obtained from deep-sea cores drawn from the depths of the Pacific. For more than seven hundred thousand years, the world's climate had oscillated constantly from cold to warm and back again. Another cold cycle would, in the fullness of time, tip earth back into another glaciation. Another, increasingly vociferous, school of thought argued that humanity's promiscuous use of fossil fuels and other pollutants was triggering unprecedented global warming. Could the Ice Age end precipitously within a few centuries with the wholesale melting of arctic and antarctic ice sheets, dramatic rises in world sea levels, and widespread desertification of now fertile lands—at a time when the world's population would be several times larger than today?

For the first time we faced the reality of a global climate: Atmosphere-ocean interactions in one part of the world have drastic effects on densely packed human populations thousands of kilometers away. Furthermore, people's fates depended not on generational memory but on policy decisions made by governments that lacked the resources to feed their peoples in times of flood or drought or when

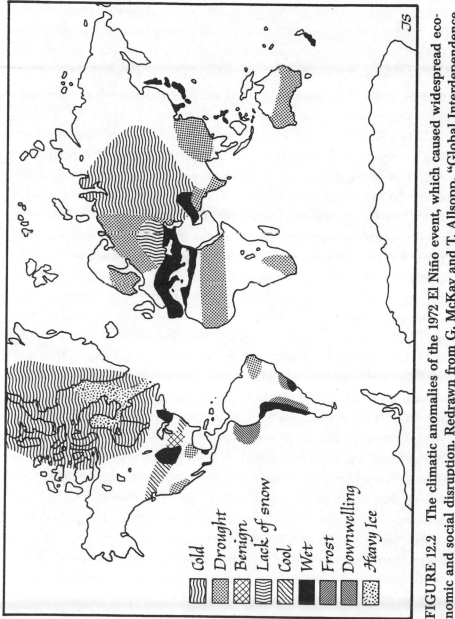

FIGURE 12.2 The climatic anomalies of the 1972 El Niño event, which caused widespread economic and social disruption. Redrawn from G. McKay and T. Allsopp, "Global Interdependence of the Climate of 1972," *Proceedings of the Mexican Geophysical Union Symposium on Living with Climate Change*, Mexico City, May 1976, 79–86; reproduced with permission.

growing populations outstripped the land's ability to support them. El Niños and the entire issue of global climate change and human vulnerability assumed an urgent scientific importance.

Although the 1972–1973 ENSO triggered much basic research, El Niño scientists labored in relative obscurity for a decade. Then the intense 1982–1983 event dwarfed its predecessor while catching many meteorologists by surprise.

Like 1972, 1982–1983 was a year of weather anomalies, most of which were, rightly or wrongly, attributed to El Niño. This particular event developed differently from its immediate predecessors. Unlike recent ENSOs that had developed off the Peruvian coast in April and then spread westward into the Pacific, the 1982 event began with unusually warm sea-surface temperatures in the central and eastern tropical Pacific between June and August and a rise in sea-level barometric pressure in the western Pacific. The warm temperatures moved eastward toward the South American coast, but west-flowing winds still blew as usual along the Peruvian shoreline as late as August. In November, much later in the year than usual, El Niño conditions developed in earnest, catching forecasters by surprise. One cannot blame the scientists. They lacked the comprehensive satellite monitoring systems that would become available in the 1990s, allowing forecasters to track sea-surface temperatures far offshore and to create computer models of a developing El Niño from birth to death.

The resulting El Niño caused the warmest winter in the eastern United States in a quarter-century and the quietest Atlantic hurricane season the twentieth century had yet seen. But southern California suffered a winter of powerful southeasterly storms that pelted the coastal regions with torrential downpours, flooded freeways, and sent more than thirty houses cascading down muddy hillsides. Huge swells swept away piers, damaged beach properties from Santa Barbara to San Diego, and spawned tornadoes. The tempests caused $1.8 billion of damage in the mountain and western states alone. El Niño effectively stopped or reduced coastal upwelling along the California coast. The squid fishery was devastated. Salmon runs off Washington State moved northward into cooler Canadian waters. The Gulf of Mexico states suffered from flooding and record rainfall. Drought reduced corn and soybean production in northern and central states.

One of the 1983 storms caught me unawares off the inhospitable central California coast. We had set sail northward from Santa Barbara to San Francisco three days before on the wings of a northeaster blowing off the land. Later we rejoiced when the wind swung to the southeast as what NOAA Weather Radio called "a weak cold front" made its way across the coast. The forecasters having called it a slow-moving system, we thought we had time to slip into the Golden Gate with a fair wind astern. But the front moved through rapidly, sprinkling us with heavy rain. Then the wind shifted to the northwest within minutes: twenty-five, thirty, forty knots dead on the nose. Fortunately, we were thirty kilometers off Big Sur, so we shortened sail and lay to the storm comfortably under heavily reefed sails. Two days passed before the winds following the "weak cold front" calmed down and we could resume our voyage.

The 1982–1983 ENSO and related weather anomalies set records all around the world. Severe droughts affected a vast area of the Soviet Union in 1982, including the country's most important agricultural regions. Cereal production fell to 205 million metric tons, compared with 237.4 million in a good rainfall year like 1978. Meanwhile, Britain experienced the fourth-warmest January of the century in 1983 and, in July, the highest monthly average temperature value (19.2 degrees Celsius) in over three hundred years of record-keeping. By midsummer some areas of Spain had experienced three years of drought; 174 towns and villages had little or no water, and 1.3 million people were affected.

In the Pacific, French Polynesia suffered through six unseasonal tropical cyclones that cast dozens of yachts ashore, sank fishing boats, and destroyed thousands of coconut palms. Hawaii endured a

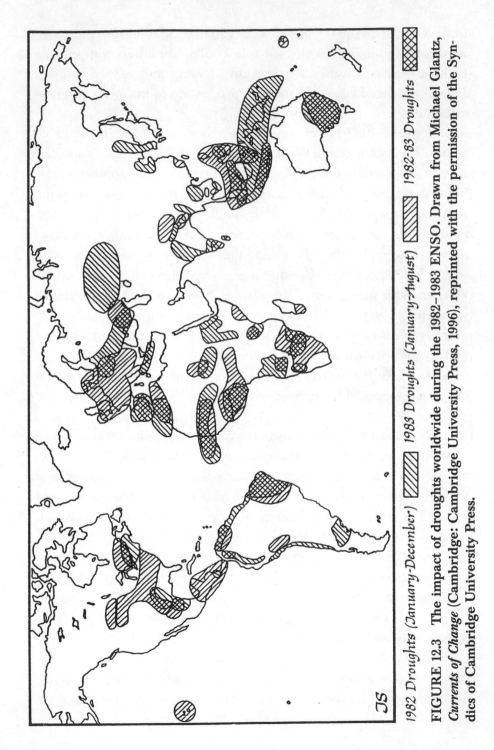

1982 Droughts (January–December) [diagonal hatch] *1983 Droughts (January–August)* [diagonal hatch] *1982–83 Droughts* [crosshatch]

FIGURE 12.3   The impact of droughts worldwide during the 1982–1983 ENSO. Drawn from Michael Glantz, *Currents of Change* **(Cambridge: Cambridge University Press, 1996), reprinted with the permission of the Syndics of Cambridge University Press.**

drought and a hurricane that caused $230 million damage. Wherever weather anomalies occurred, the poor suffered. Six hundred people, mainly city slum dwellers, died in the savage floods that hit coastal Ecuador and northern Peru. Storms and floods devastated entire cities and the coastal infrastructure of roads, railroads, bridges, and irrigation schemes.

Southern Africa had irregular rains and much drier weather than usual. The journalist Allister Sparks traveled from Johannesburg to Cape Town in South Africa's famed "Blue Train" and witnessed the severe drought caused by the El Niño at firsthand. As the luxurious train crossed the fertile corn belt of the western Transvaal and Orange Free State, he saw a landscape baked "to parchment," the withered fields and stubble suffering through the worst drought since records began in the 1840s. Arid lands gave way to lush green meadows and full rivers 150 kilometers from Cape Town, which itself had received 1,900 millimeters of winter rainfall in five months, double its annual average.

South African grain production was off by 44 percent in 1982, and 46 percent in 1983. The nation, normally a corn exporter, had to import one and a half million tons of maize from the United States. The black population suffered worst, partly because apartheid's relocation schemes had forcibly resettled three million Africans in segregated "homelands" over the previous twenty years. A deadly combination of overcrowding and drought caused traditional subsistence agriculture to break down. More than one-third of rural children suffered from malnutrition, and farmers ate the seed grain sent them by relief agencies. Worst hit was the Ciskei homeland in eastern Cape Province, where the government had doubled the population in ten years through compulsory resettlement. In what one opposition member of Parliament described as a "shattered territory," half the sheep and cattle had perished.

The El Niño brought great suffering to Northeast Brazil. The "Nordeste" is always drought-prone, with high rainfall near the coast. Then a drier belt inland between the coast and the tropical forests of the Amazon Basin. When the droughts arrive, they hit hardest in the poverty-ridden semi-arid zone, which receives between 300 and 800 millimeters of rain a year. Brazilian scientists call this the "drought polygon"—a low Southern Oscillation Index and ENSO conditions often coinciding with low rainfall years. In 1983, 88 percent of the Nordeste suffered from severe drought. Agricultural production declined by 16 percent, and fourteen million people were affected by the dry conditions. Small landowners and subsistence farmers were the most vulnerable. Many of them lost their entire crop and fell deep into debt when food prices, controlled by corrupt officials and profiteers, rose by as much as 300 percent. As drought assistance, 2.8 million subsistence farmers were given low-wage employment. Meanwhile, southern Brazil saw record rainfall and hundred-year floods.

On the other side of the world, severe drought settled over Southeast Asia. It dried out the tropical forests of the Kalimantan region of eastern Borneo, which usually received more than 2,500 millimeters of rainfall annually. The dry conditions caused evergreen trees to shed. Dry litter accumulated on the forest floor. Between August 1982 and May 1983, oven-hot winds fanned huge forest fires ignited by farmers burning off their land. Rising population densities caused by accelerated settlement programs and migration had created insatiable demand for new farming tracts. Inevitably, the farmers' fires burned out of control and moved into virgin forest. The dry conditions had lowered the water table, killing shallow-rooted trees and turning freshwater swamps into tracts of tinder-dry peat. Three and a half million hectares of surface and underground fires raged unabated for almost three months in early 1983. Dense clouds of wood smoke hung over the island and drifted as far west as Singapore, fifteen hundred kilometers away. Airports closed because of the smoke and haze. The Kalimantan port of Balikpapan closed for weeks because ships could not navigate the entrance.

The Borneo fires of 1983 were one of the worst environmental disasters of the past century. Prophetically, the scientist Jean-Paul Malingreau wrote two years later: "If . . . intensified land use increases the risk of fire in tropical ecosystems, the impact of future droughts is likely to be even worse."[1]

The litany of weather anomalies and disasters seems unending. Australia suffered through a drought that affected 60 percent of its 67,000 farms. Crop production fell by 31 percent. Even irrigated farms growing cotton and rice suffered from water shortages. Wind erosion blasted the dry grain fields and pastures. A dust storm in February 1983 carried at least 150,000 metric tons of soil from farms in Victoria into Melbourne and far offshore. Bush fires in southeast South Australia and Victoria consumed 500,000 hectares, killing 72 people and more than 300,000 animals. Some of the burned lands lost up to 48 metric tons of soil per hectare in the heavy rains that eventually broke the drought. The $3 billion price tag encouraged official efforts to forecast El Niños.

Estimates are always imprecise, but the global damage wrought by the 1982–1983 El Niño and related climatic anomalies cost over $13 billion. More than two thousand people died. The "Mother of El Niños" caught everyone's attention, not just the scientists'.

The 1982–1983 El Niño generated intense research into ENSO. In early 1986, Mark Cane and Stephen Zebiak of Columbia University's Lamont-Doherty Earth Observatory developed a simple model of interacting atmospheric and ocean conditions in the tropical Pacific and boldly forecast that an El Niño was imminent. Many of their colleagues were aghast at their temerity, but when a moderate ENSO did develop in 1986–1987, other researchers were encouraged to follow suit. By this time the study of global climate was advanced enough to have developed the first regional early warning systems for impending food shortages and famines. The National Meteorological Services Agency started issuing seasonal forecasts.

The agency drew the attention of governments and others to developing drought conditions in northeast Africa early in 1987. A low Southern Oscillation cycle and warm sea temperature anomalies over the southern Atlantic and Indian Oceans caused the summer rains to fail in Ethiopia. The Ethiopian government responded by encouraging farmers to plant as much as possible during the short rains from mid-February to mid-May. The resulting bumper crops saved thousands of lives when the longer summer rains failed. At the same time, international relief agencies shipped in emergency supplies ahead of time. Although most of the summer crop was lost and many cattle died, not a single human perished from starvation, despite political unrest and warfare in the region.

In early 1991 Cane and Zebiak once again issued an El Niño forecast. The U.S. National Meteorological Center also predicted the same episode, which began in 1991 and continued with three distinct sea temperature warmings over the next four years—either the longest El Niño in 130 years or three separate episodes in rapid succession. Once again, the Christmas Child proved unpredictable. Forecasters learned the hard way that no ENSO event is like its predecessors. However, there were some notable successes. The province of Ceará in Brazil's Nordeste created the Meteorological Foundation specifically to predict El Niño–connected droughts. The foundation's forecasters took heed of the government forecasts in the United States and issued a severe drought warning in 1991. The provincial governor traveled widely through Ceará, encouraging farmers to switch to crops that would grow faster and mature sooner in a short, poor rainy season. Thirty percent less rain fell, but grain production fell only 15 percent, whereas in the 1987 drought it had declined by 82 percent. The foundation issued two more successful drought forecasts and enjoyed high credibility. But when it wrongly predicted El Niños during the early 1990s, the fickle public became disillusioned, unaware that these particular episodes had been some of the most unpredictable on record.

Even before a wave of missed forecasts in the early 1990s, scientists had few illusions as to the reliability of their long-range predictions. Governments, politicians, and the public at large tend to take climatic predictions at face value, as if they were medical diagnoses, while ignoring the forecasters' carefully worded qualifiers. The 1997–1998 El Niño is a case in point. October and November saw sunny skies and perfect temperatures. "Where's El Niño?" Californians laughingly asked. The scientists counseled patience, but even they began to wonder whether their forecast was wrong. In January and February the forecasters had the last laugh when the Christmas Child brought furious storms to the coast. Nevertheless, a meteorologist's credibility is always on the line. Sir Gilbert Walker remarked as long ago as 1935: "It is the occasional failures of a government department that are remembered."[2]

In early 1997 the oceanographers' computers picked up a mass of unusually warm water in the southwestern Pacific. The new International Research Institute for Climate Prediction (formed jointly by Columbia University's Lamont-Doherty Earth Observatory, the Scripps Institution of Oceanography, and the National Oceanic and Atmospheric Administration) launched a coordinated effort to monitor the phenomenon and to predict the effects of a rapidly forming El Niño on widely separated parts of the world. The new El Niño was the most thoroughly studied climatic event in history.

Researchers watched as the warm water mass grew and spread rapidly through the equatorial Pacific. The warm ocean pumped heat and moisture into the atmosphere. By August sea temperatures off South America were already as warm as they had been during the great El Niño of fifteen years earlier. For the first time, forecasters could use their models and the experience of the 1982–1983 ENSO to issue an accurate long-term forecast. They predicted higher than normal rainfall for most of California, the Southwest, and southern states, with drier conditions in the Midwest east of the Mississippi River. A warmer winter was likely in the upper Midwest and in the Northeast. Over the northern summer months of 1997, the fledgling El Niño developed just as the forecasters predicted, as severe drought descended on Australia and Indonesia. The global scenario was eerily familiar.

On November 8, 1997, a night of heavy rain over the banana-producing town of Santa Rosa in southern Ecuador caused rivers to burst their banks. Within minutes, over three thousand people were homeless. Banana and cocoa plantations were inundated, and shrimp farms decimated. The inhabitants of Santa Rosa were furious. They pointed out that the surprise 1982 El Niño had cost Ecuador $165 million in damage to farms, housing, industry, the country's infrastructure, and fisheries. Why, then, was the government unprepared for disaster when it had had months of warning of a strong ENSO event? The authorities responded that since Santa Rosa had been spared damage in 1982, preventive measures were directed elsewhere. They also pointed out that times had changed. For example, Ecuador had only 35,000 hectares of shrimp farms in 1982. In 1997 there were 180,000 hectares, many of them hacked out from coastal mangrove swamps that once formed a natural barrier against flooding. Now rising river waters easily knocked down the retaining walls around shrimp hatcheries. Predictably, the interior minister announced that "we will have to manage with what we have"—meaning, it was every family for itself. Local authorities tried, with only limited success, to dissuade people from blocking ravines and drainage ditches with garbage. El Niños are an issue that will not go away in a country where political cynicism runs deep. Public opinion polls found that Ecuadorians were more concerned about El Niño than about the membership of the National Assembly.

The Peruvian coast also suffered violent rainfall that flooded thousands of hectares of agricultural land. The city of Piura had twelve separate days with at least half its normal annual rainfall. Another town, Talara, received five times its annual rainfall in a single day. In Trujillo, Peru's third-largest city, floodwaters swept away poor neighborhoods and slums. Muddy water engulfed a cemetery and emptied 123 graves. The survivors watched as their recently departed relations floated downstream like ghosts in their funerary suits and Sunday dresses. Flooding, mudslides, and disease killed more than eighty people. The president of Peru and his entourage descended on Trujillo, bringing emergency shelters designed by a New Jersey company. Cynical locals noted that political officials declined to enter damaged houses on account of the smell left by receding floodwaters.

By December, El Niño's warm waters were pulsing every thirty to fifty days, contracting by 10 to 15 percent as their heat dissipated into the atmosphere, then being replenished by strong winds from the central Pacific. The Topex/Poseidon satellites gave scientists a continuous picture of El Niño's "plate" of warm water, which by this time covered an area greater than the United States and had heated a blanket of air that extended from the eastern Pacific across Central America into the Atlantic from Cuba to Libya. The warm air shouldered aside normal seasonal weather patterns to impose its own.

February 1998 saw a change in the Northern Hemisphere jet stream. Warm tropical air, pulled into the north, brought above-normal temperatures to North America, Europe, and eastern Asia. Land temperatures in the Northern Hemisphere reached their highest levels since 1950. Globally, the combined air temperature over land and sea surface for February was 0.75 degrees Celsius above normal, breaking the record for the highest departure from the 1961–1990 mean for any month back to 1856. February 1998 was not only the warmest February since records began, but it also broke the reading by more than ever before.

Brazil's Nordeste went through a severe drought, as predicted. As always, ten million poor and landless bore the brunt of it, as a slow-moving federal government neglected them. In Pernambuco state, local political bosses and wealthy landowners have developed a "drought industry" that controls relief supplies and makes the poor dependent on emergency supplies as a way of controlling their votes. In the past the underprivileged simply endured this arrangement. In 1998, having been forewarned of the drought and backed by the Landless Workers' Movement (MST), they rose in hunger and indignation. Angry crowds looted supermarkets, food trucks, and warehouses in well-organized protests. Even some truck drivers and warehouse managers sympathized with the hungry, looting mobs.

The poor had rebeled against a system in which powerful landowners control water and land with private armies. They depend on cheap labor for their wealth, which comes from old-fashioned farming methods, not from industrial-scale agriculture that can reclaim devastated fields in semi-arid areas. The Landless Workers' Movement has support from many Brazilians, even from Catholic priests, who consider the right to human life a fundamental ethical value. Brazil's National Council of Bishops has allowed that looting may be ethical if survival is at stake.

The drought came in a presidential election year. Incumbent President Fernando Henrique Cardoso was up for reelection. He attacked MST for organizing looting as a new industry that preyed on the poor during droughts and announced a $500 million aid program for the affected areas that included food baskets, new irrigation schemes, a work relief program, and ambitious plans for job training and literacy programs for at least one million people. The president's poll ratings soared, and lawlessness and looting subsided somewhat, but unrest still simmered through the Nordeste. As of this writing, the landless of Pernambuco state were living off food baskets through the dry months of July and August 1998, waiting for the new rainy season and wondering what would happen when the food baskets ran out.

Global forecasts and satellites are a two-edged sword. They provide information for governments, but also for those they govern. Brazil's landed elites do not have a monopoly on weather forecasts. The poor are better informed as well, and as Egyptian pharaohs learned the hard way, information is potential power. If the Brazilian government does not act decisively on basic issues of land reform and other structural changes, the next drought will inevitably bring serious political and economic disorder. Famine resulting from drought is as much a social disaster as an environmental one.

On the other side of the Pacific, Philippine rice paddies, usually under nearly two meters of water by mid-June, were a vast expanse of red cracked earth covered with tufts of dying rice. The farmers' irrigation canals were dry and clogged with weeds. Temperatures soared above thirty-seven degrees Celsius for weeks on end. This was the second year of a drought that had already cost more than $20 billion in lost crops, caused several hundred deaths, and led to serious food shortages. Some rain had finally arrived, but the worst dry spell to hit the region in more than four decades persisted in many areas. Many farmers turned their rice plots into fish ponds, but the water was too hot and the fish died. Others grew swamp cabbage, a form of spinach that has long served as a famine food for rice farmers. They did everything they could to keep their water buffalo alive, for when the rains arrive they need them to plow their land. Thousands of farmers moved to the cities, only to find there was no work.

The same drought found Indonesia wrestling with uncontrollable forest fires. Like the Borneo fires of fifteen years earlier, these began when farmers and plantation owners set fire to uncleared land on the assumption that monsoon rains would soon extinguish the flames. The rains never came. The flames roared out of control through the dry forest, killing millions of trees and their root systems. Thick clouds of smoke, ash, and haze blanketed Southeast Asia, disrupted air travel, and caused at least one airline crash that killed 234 people. Smog levels rose to an unprecedented level four times that of a second-stage smog alert during a hot summer day in Los Angeles during the 1970s. Businesses and schools closed. People collapsed in Jakarta's streets, and birds dropped out of the sky. The forest fires caused economic losses of at least $1.3 billion. More than seven and a half million people in fifteen Indonesian provinces faced serious food shortages and required outside assistance.

The effects of the El Niño drought will continue in Southeast Asia even after the rains return to normal. The great forest fires have stripped millions of hectares of their natural protective covering, raising the specter of widespread soil erosion and destruction of once-fertile land. At least half of the 900 million hectares of rice farmland in the Philippines are subject to flooding—and floods destroy not only crops but human lives as well.

These disasters unfolded even though the 1997–1998 El Niño was tracked from the moment it first appeared as a growing mass of warm water in the tropical Pacific. Regular forecasts and synopses appeared on the World Wide Web, where the World Meteorological Organization's *El Niño Updates* summarized current developments and made long-range predictions on the basis of sea-surface temperature, local observations, and historical data on the effects of ENSO episodes. In February 1998 the World Meteorological Organization reported:

> Warm episode conditions are expected to continue February through April and to weaken during May-July. Drier-than-normal conditions are expected over Indonesia, northern South America and parts of southern Africa during the next few months. Wetter-than-normal conditions should continue over the central and eastern equatorial Pacific, along the coasts of Ecuador and northern Peru, and over southeastern South America. Also, increased storminess and wetter-than-normal conditions are expected to continue over California and the southern third of the United States. Warmer-than-normal conditions will persist over much of central North America.[3]

ENSO droughts and floods mean famine. Early warning programs went into high gear, including the U.S. Agency for International Development's Famine Early Warning System (FEWS), which gave advance notice of drought and potential crop failure in Northeast Brazil, southern Africa, and Southeast Asia. The reports were accessible to everyone and could be downloaded from the Internet. Said a report for African countries issued in mid-1997: "NOAA has now issued an ENSO advisory that a warm event is developing. Having started in the April-May period and having developed quite rapidly, this may be one of the stronger episodes. FEWS and other groups are monitoring this event carefully to track its development and determine its likely effect on weather and crops."[4] The report included maps that showed the areas most at risk. On the basis of information about the impending El Niño, Ecuador borrowed $180 million from the World Bank to offset the effects of abnormal weather on the national economy; Peru borrowed $250 million. Agricultural officials in Cuba ordered an early start to the sugar harvest to avoid potentially damaging storms. International relief agencies stockpiled relief supplies.

The bill for the 1997–1998 El Niño is not yet in, but the cost will vastly exceed the $13 billion of its 1983 predecessor, not necessarily because the droughts were more severe, the rains and storms more savage, or the fires so widespread, but simply because there are now so many more people in the world, many of them crowded into cities and urban slums. The same people are also much more aware of the hazards (and benefits) of El Niño. Even the poor receive the forecasts and sometimes act on them, when they are able. However, this better-informed public has much higher expectations of governments and international relief agencies than even a generation ago. The social unrest caused by the two greatest El Niños of the twentieth century made few headlines, but we can confidently expect civil disorder, food riots, and looting to become a familiar part of major ENSO events. As the world's population continues to grow, its long-term environmental debts continue to accumulate.

Between March and August 1998, the pool of abnormally warm water in the Pacific shrank steadily. Large areas of the eastern Pacific enjoyed near-normal temperatures as the greatest El Niño of the twentieth century came to a close. The Southern Oscillation pendulum began a swing toward a predicted La Niña in 1999. In the meantime, millions of people traded their El Niño stories and jokes. An Oregon company marketed a battery-operated animal toy with a badge saying, "My name is El Niño. You can blame me for everything." And in the ultimate compliment to the world's maverick weather machine, an elementary school student in Lakewood, California, when asked what season it was, replied without hesitation: "El Niño."

CHAPTER THIRTEEN

The Fate of Civilizations

The statesman's task is to hear God's footsteps marching through history, and to try and catch on to His coattails as He marches past.

—*Otto von Bismarck*

We are *Homo sapiens sapiens,* the clever people, animals capable of logical reasoning, subtlety, and self-understanding. Born in the crucible of the Ice Age, we have always lived in a world of constant climatic swings. As the physiologist Jared Diamond has recently pointed out, the striking divergences between the histories of people living in widely separated parts of the world are due not to innate differences between the people but to variations in their environments. We have been successful at adapting to such diversity because of our unique ingenuity and inventiveness. Writes Diamond: "Without human inventiveness, all of us today would still be cutting our meat with stone tools and eating it raw, like our ancestors of a million years ago."[1] To which one could add that without these qualities, forged during the Ice Age, we would not have been able to navigate the unpredictable climatic rapids of the past ten thousand years.

Until ten millennia ago, all humanity lived by hunting animals large and small and by gathering wild plant foods. For most of this time, the global population was nearly stable, growing at a rate of just 0.0015 percent annually as gradual technological improvements enhanced humans' ability to live off environments that could support much less than one person per square kilometer. For hundreds of thousands of years, there was plenty of room for people to move around—to come together in plentiful months and disperse in times of scarcity. They always had mechanisms to help one another, as well as the space to disperse when famine threatened. Altruism, common interest, and reciprocity were powerful forces for survival in egalitarian forager societies, where leadership depended on experience and flexibility, as it does today among the Kalahari San with their *hxaro* networks.

Ten thousand years ago, the world was close to the limits of its ability to support people living by foraging alone. At the time, an estimated ten million people lived on earth. Then a sudden flip in the North Atlantic circulation and plunging global temperatures brought severe drought to the well-watered valleys of southwestern Asia. As nut harvests failed and food supplies dwindled, the people used their knowledge of plant germination to cultivate wild cereal grasses, which happened to abound in this region. No more than one or two generations later, the foragers were farmers and human life had changed forever.

The new economies were highly successful at solving hunger problems in the short term. But they created frightening challenges on a millennial scale as human populations rose far faster than ever before. Without continuous human innovation, this growth would have once again outstripped the ability of the world's environments to support more people.

Southwestern Asia had great potential for agriculture and animal husbandry because of its varied environments, indigenous wild cereal grasses, and potentially domesticable animals. Human population growth accelerated to 0.1 percent annually as agriculture spread through Europe and the Mediterranean Basin; it developed independently in China and Southeast Asia, then later in the Americas. By the time the first civilizations developed, in Egypt and Mesopotamia in 3100 B.C., there were as many as twenty million human beings on earth, many living in the fertile regions where the first urban societies appeared. At this point, basic subsistence agriculture gave way to more intensified cultivation that provided higher crop yields per hectare, sometimes even two or three harvests annually. Such farming required careful organization and leadership and was successful in Egypt and Mesopotamia, where power and authority were vested in divine kings.

The early preindustrial civilizations grew over many centuries. By 2000 B.C., a vast network of trade routes and political interconnections linked much of the eastern Mediterranean world. By the time of Christ, the discovery of the monsoon winds had linked that world with East Africa, India, Southeast Asia, and China. The world's population still expanded rapidly through classical times and into the modern era, despite the depopulation caused among Native American populations by unfamiliar epidemic diseases introduced by Europeans. By A.D. 1800, when the Industrial Revolution began, the world's population had reached one billion people.

The Industrial Revolution changed the face of the world with technologies that improved communications and medical care, agricultural production, and humanity's ability to transform the world's environments. Millions of people took steamships to distant lands, whether as colonists, indentured laborers, or slaves. Wherever the Industrial Revolution took hold or Europeans colonized, exponential population growth followed. There are six billion of us today, and the increase continues. About 240,000 more people are born than die each day. At our present growth, there will be twice as many of us in forty years, despite near zero population growth in many parts of the developed world. We are now dangerously close to the technological limits of what we can achieve to feed everyone on this bountiful earth. No number of green revolutions, desert irrigation schemes, or fish farms can satisfy the food needs of such expanding numbers. Nor, as Carl Sagan recently pointed out, can we ship 240,000 people to other planets in outer space every day.

Since ten thousand years ago, humanity has grappled with the complex problem of feeding itself in the face of growing numbers of people and unpredictable climate change. Fortunately, we are unique among animals in our ability to transmit, through language, knowledge and experience from one generation to the next, so each benefits from the cumulative experience of earlier times. For ten millennia, the ability of some humans to become leaders, and of all of us to cooperate with one another, combined with ever more effective technology, has allowed us to cheat the realities of a finite environment and to sustain a growing humanity. The massive population increases of the past two centuries are putting these abilities to their greatest test.

The issue is sustainability. In 1798 the British clergyman Thomas Robert Malthus drew attention to the dangers of population growth in his famous *Essay on the Principle of Population*. He argued that human population has a natural tendency to increase faster than the means of subsistence, and he made a powerful argument for some form of population control. Malthus's judgments were colored by the regular subsistence crises that had beset Europe's poor during the Little Ice Age. But he was right in drawing attention to the delicate balance between human populations and the ability of any particular environment to support them indefinitely—what modern-day scientists call the carrying capacity of the land. This is sustainability: indefinite support without degrading the environment and causing a lower carrying capacity in the future. The long-term equation of people and sustainability allows for such variables as seasonal rainfall, droughts, and El Niños. The danger comes when the equation goes out of balance as a result of exponential population growth, or when shortsighted human behavior, such as unbridled forest clearance or uncontrolled grazing, causes soil erosion or incipient desertification.

Such behavior is nothing new. We know that much of southwestern Asia was deforested within a few millennia of farming activity.

Years ago I climbed the reconstructed Sumerian ziggurat pyramid at biblical Ur, brilliantly excavated by the Englishman Leonard Woolley in the 1920s and 1930s. I gazed out over a desolate, sandy wilderness and remembered Woolley's accounts of a prosperous city teeming with life and movement. Four and a half millennia ago, ocean-going ships docked at Ur's quays. The waters of the Euphrates nourished thousands of hectares of lush cropland. Ur became one of the great cities of the ancient world. Then the river changed course, salt levels rose owing to intensive irrigation of poorly drained soils, and the kings vanished. Ur became a desolation, in part because humans destroyed the city's environment.

Carrying capacity is one of the keys to global food security. But the concept makes many social scientists skeptical. They point out that people, unlike animals, can get around environmental limits by using human labor or technology to intensify agriculture and grow more food, or to herd more animals. In other words, carrying capacity is not static, but flexible. As the Danish economist Ester Boserup pointed out in 1965, population growth can cause people to intensify their farming with irrigation and other productivity enhancements. The record of history is that this is exactly what they did.

A significant body of thought, headed by the economist Julian Simon, asserts that carrying capacity is *infinitely* flexible—that in the modern world, technological improvements will always keep productivity running well ahead of population increases. As proof, Simon points to the falling real prices of food, minerals, and other commodities throughout the twentieth century. Only time will tell if this proposition is true over the long term. We archaeologists know that it has never been true in the past; we make our living off history's counter examples.

To subsistence farmers, human existence and the environment have always been as one, each supporting the other, with understood, but unspoken, limits. The living world is an integral part of a much larger cosmos, where the material and supernatural flow imperceptibly one into the other. In the past, village farmers lived in tight-knit communities that persisted for generations, their lives bound by an endless cycle of land clearance, planting, and harvest. Their ancestors, those who went before, were the guardians of the land, and in due time the living became ancestors as new generations inherited the land. Ancestors interceded with the unpredictable forces that brought rain and drought, storm and flood. Even in the most egalitarian farming societies, some individuals acquired unusual secular powers through their ability to communicate with the ancestors and with the forces of nature, or through their genealogical ties to a mythic hero. They mediated disputes, presided over planting and harvest rituals, and used people's loyalty to harness rivers and irrigate the land.

In 3000 B.C., the moment when the first civilizations appeared, crowded urban and river valley populations also became increasingly vulnerable to short-term climatic change. El Niños became significant players in history because a growing proportion of the world's population could not move away from drought, flood, or famine.

This moment was also when human beings became gods for the first time. Until the birth of Western science and the Industrial Revolution, every human society assumed that sudden climate change was willed by the gods. Poseidon was angered at Odysseus and sank his raft with a violent hurricane. The Lord brought plagues and hunger down on Egypt. The Aztec rain god Tlaloc brought fertility or drought to the earth. It was no coincidence that the rulers of the first civilizations claimed ancestry from the divine or power over terrible storms. This was the source of the authority that allowed them to coerce people into laboring for the common good. They also created new levels of social inequality that created the ever growing gap between the rich and the poor, the rulers and the ruled.

Ultimately, leadership, and the sanction for the inequality that accompanied it, came from men and women who were living deities on earth. Right from the beginning of civilization, a vast social chasm separated the ruler and the tiny number of wealthy or affluent families from the thousands of urban and rural poor. The anonymous hands of the commoner and the slave supported the magnificence of pharaohs' courts and Maya cities and bore the brunt of the gods' wrath. The ravages of famine and El Niños have fallen hardest on the underprivileged for more than five thousand years. When climate-induced hunger caused the commoners' hands to cease work, tribute payments to falter, or social unrest to erupt, ancient civilizations could collapse, because divine leadership had feet of clay and the rulers lost their infallibility.

The great nineteenth-century German statesman Otto von Bismarck once observed that humanity floats on a "Stream of Time," an ever flowing river that carries us along through calm stretches and rapids. We can neither create nor direct the Stream, but can merely do our best to navigate it. The vagaries of climate change have shaped the quiet reaches and waterfalls of this endless stream since human history began. Until recently, we lacked the scientific tools to appreciate just how profoundly short-term climate change has affected the rise and fall of civilizations.

We tend to assume that the great El Niños of the late twentieth century are unique events, unparalleled in history. Like our forebears, we still look at the vagaries of global climate through the narrow blinkers of human memory and the experience of a few generations. This is hardly surprising, since accurate meteorological records are a mere 150 years long in Europe and much shorter in many other parts of the world. Even rudimentary studies of global climate date back little more than three centuries. Thanks to highly sophisticated paleo-climatic research and new historical studies, we are discovering massive El Niños of the recent and remote past, like the great ENSO event of 1792 that brought drought to many parts of the world and another turbulent episode that nearly decimated the Chimu state in coastal Peru during the twelfth century. Severe climatic disruptions rippled across the Mediterranean world in about 1200 B.C. Our present weather may be warmer than usual, but its unpredictability is nothing unusual.

Since 1983 El Niños have assumed near-mythic proportions in the public eye. They have been blamed for everything from chronic traffic jams to famine-causing droughts, epidemics, and delayed PGA golf tournaments. To grant them such omnipotent powers is to overstate the case. Although for five thousand years we have become increasingly vulnerable to severe El Niño episodes, they have never, on their own, caused the collapse of any civilization. But they have acted on many occasions as a knockout blow.

The Moche civilization of coastal Peru is a case in point. Moche lords inherited a long tradition of river valley irrigation agriculture that harnessed spring mountain runoff from the Andes Mountains. They refined ancient river canals, reservoirs, and field systems that supported valley populations in the thousands, far more than the natural carrying capacity of their river floodplains. Like other preindustrial civilizations, Moche society was a social pyramid, dominated by a few powerful and wealthy lords who competed with one another from atop their great adobe pyramids built by their laboring subjects. Because the Moche did not possess writing, we cannot decipher their religious beliefs. But the highly standardized and flamboyant regalia worn by the lords of Sipán hint at a long-lived, very structured ideology that left little room for innovation from Moche rulers. Inflexible, despotic lords like these, who probably considered themselves living gods and infallible, were ill equipped to deal with the famines and accelerating social stress brought on by long drought cycles, unpredictable floods, and periodic strong El Niños. In the sixth century A.D., they survived one drought and catastrophic cycle of earthquakes and El Niño, which caused them to move their capital far upstream to maximize floodwaters and crop yields. There they minimized water loss in drought years. But they could not withstand a later El Niño or series of ENSO events. A combination of rigid governance, drought, social stress, and El Niños toppled centuries of prosperous rule.

The same pattern of rapid rise, efflorescence, and sudden collapse epitomized many other preindustrial civilizations, among them the Classic Maya, whose lords were as exploitative and myopic as the Moche warrior-priests. A generation ago we would have attributed these implosions to complex social and economic or environmental factors. Today we realize that sudden climate change was a key player in history, especially as a final blow that undermines generations of escalating stress caused by poor leadership, growing population densities, and environmental degradation—to mention only a few of the cumulative effects of living out of balance with the natural carrying capacity of one's homeland.

Few ancient leaders understood that their divinity was a pretense. Remote and despotic, the Old Kingdom pharaohs considered themselves infallible masters of the Nile. When El Niño proved them wrong with a generation of drought, their authority withered. Disillusioned local leaders and nomarchs, with closer ties to the land, looked after their own people and ignored their divine ruler. The later Middle Kingdom pharaohs, well aware of this lesson and themselves descended from nomarchs, never preached infallibility. They became administrators and shepherds of their people, with a deep sense of their responsibilities. These pharaohs endured another eighteen centuries because they were realists, determined that no drought could ever totally undermine what they had achieved. They learned the hard way to exercise tighter social control over their domains and did all they could to increase the natural carrying capacity of the Nile Valley for a growing population through large-scale irrigation and improved technology.[2]

History teaches us that the best leaders were prepared to learn from experience and hard lessons. Flexibility and far-sightedness are rare leadership qualities, especially in societies, shackled by rigid secular or religious ideologies, where drought and flood are attributed to the whim of the gods. Such civilizations survived in the short term because doctrines like the Egyptian *ma'at* or the fictional genealogies of the Maya validated social inequality. Thousands of laborers cultivated and harvested grain, built irrigation systems, maintained field systems, and erected enormous pyramids. These thousands of hands were society's bastion against sudden flood or drought, against capricious El Niños. However, such kingship required extraordinary qualities of responsibility and long-term vision, for it was the commoners who fed and maintained the social order.

Faith in leadership was vital in societies where everything functioned through the application of organized human power. This leadership could be benign, humane, and wise, like that of some Middle Kingdom pharaohs, or it could be ideologically inflexible, self-absorbed, and exploitative, like that of the Maya lords. Every leader, good or bad, rationalized his or her deeds as a function of a special relationship with the spiritual world. But as the Andean lord Naymlap found to his cost, a ruler who did not deliver food and safety to his people was in grave danger, especially when his political antennae failed to identify mounting social stresses around him. Rulers today, from American presidents to Soviet leaders to uncrowned monarchs in the Third World, face the same challenge.

There was no such thing as a global perspective in a world where Central America, Tahiti, or Australia was as remote as the moon is today, nor was one needed. Today the responsibilities of a pharaoh or an Inca god-king are dwarfed by those of a modern president or prime minister in a world beset by uncontrolled population growth, deeply ingrained poverty, potential food shortages, and a new and unknown quality—humanly caused global warming. Now we contemplate the fate not of minor states or empires spread out over several ecological zones, but of global civilization.

Poverty and social inequality, urban sprawl and slums, people living on the margins of society and pushed out into marginal environments—all these problems have afflicted humanity on smaller scales since civilization began. With six billion people on earth, the classic equations of population, carrying capacity, and environmental degradation have assumed global proportions, and the lessons of history give us few precedents.

Consider the figures. About 1.3 billion, 23 percent of the global population, live in poverty, and 600 million of them are only a few hundred calories a day away from starvation. At least 400 million people eke out a living in fragile areas like Africa's Sahel, where droughts, deforestation, pollution, and other forms of environmental degradation place more and more people's livelihoods at risk.

Large-scale deforestation in the Amazon Basin has led to soil erosion and flooding. Epidemic diseases have resulted from the displacement of forest peoples by economic activity. Recently, the Indonesian government resettled nearly seven million people and cleared 4 percent of the country's forests in the process. Many of these settlers later fled back to city slums after floods, massive soil erosion, and constant crop failures. The political and social tensions arising from issues of sustainability have led to rebellions in many parts of the world.

Some of the descendants of the Classic Maya live in the Lacandón rain forest of Chiapas, where the human population has risen twenty-five-fold since 1960 while tree cover has fallen from 90 percent to 30 percent. Not only has the indigenous Maya population risen rapidly, but land-hungry peasants, Mexicans fleeing persecution, and refugees from Guatemala have settled here as well. Where 12,000 people once lived, there are now 300,000. The traditional Maya *milpa* system is undermined by the resulting environmental pressures. The people of eastern Chiapas have virtually no government services, no political power or representation, and no economic opportunities. No one should be surprised that they are rising in protest.

For every person who rebels in an obscure civil war, many more move away from their homeland. In the entire three-plus centuries between A.D. 1500 and the early 1800s, no more than three million people crossed a frontier voluntarily. Today millions of refugees are on the move, responding to shortages of food, land, and water. Ethiopia alone has over 50 million people, 30 million more than forty years ago, and faces an increase to 106 million during the next forty with an environment of severely eroded cropland and only 3 percent tree cover. Its people's options are stark: emigrate or starve. Even without the complication of possible global warming, we are in danger of a potential world food crisis. Global food production has not risen since 1990, while population has climbed by 440 million people. Many governments have assumed that farmers can simply put more acreage into production, which is not necessarily true and in many countries impossible. At the same time, tens of thousands of acres vanish annually under expanding cities.

Nor is it clear how much more can be squeezed from existing farmland. For a long time farmers relied on fertilizers to increase food production, a major cause of Peru's nineteenth-century guano boom. Between 1950 and 1989, fertilizer use increased tenfold, from 14 million tons to 146 million, a rise in consumption that tripled the world's grain harvest. Global demand for food will nearly double from its present level by 2030. How will we respond to this need? Egyptian pharaohs expanded cultivated land and stored more grain. We don't have these options. Instead, we are pinning our hopes on the methods of industrial farming and the promise of biotechnology. Even if we produce plenty of food, the political obstacles to its distribution may be severe. The twenty-first century may confront the paradox of having both more food and more food crises.

One item that is no longer a scarce commodity is information. In an era when the World Wide Web and satellites bring ENSO forecasts to households in the remotest African savanna, information is not the sole province of the powerful and wealthy. Knowledge and hunger have never coexisted easily, and the well-informed poor are sure to demand their right to food and the basics of existence.

And then there is humanly caused global warming, which most of us knew nothing about until recently. We were more worried about nuclear winter during the floods and droughts of 1982–1983, when El

Niño became a household word. Not until 1988 did the "greenhouse effect" and "global warming" become popular buzzwords. That summer was so hot that crops failed over much of the Midwest. Thousands of cattle were slaughtered for lack of feed. Barges ran aground in the Mississippi and were stranded by low water for months. Lightning storms charred millions of acres in the tinder-dry West, and huge dust clouds in Oklahoma brought back vivid memories of the Dustbowl of the 1930s. Violent rains plagued India and Bangladesh, while the Soviet Union endured a severe drought. The intense heat and economic losses prompted congressional hearings, where scientists spoke of the rising levels of carbon dioxide in the atmosphere, of ozone holes and the greenhouse effect that was contributing to humanly caused global warming. Many of them placed the blame for the record heat at the feet of humanity and its insatiable demand for fossil fuels.

Like "El Niño," "global warming" is now a household phrase. The 1980s were the warmest decade since accurate records began in the mid-nineteenth century. The 1990s have been unusually warm as well. Nine of the ten warmest years since the 1850s have occurred during the past fifteen years. The year 1997 was the most torrid on record, and 1998 continues the trend. These same years saw repeated El Niños and high sea temperatures in the tropical Pacific, which triggered extreme weather conditions in many parts of the world. The past quarter-century has been remarkable for its storminess, abrupt climatic swings, and extreme weather. The great storm of October 16, 1987, the worst since 1703, toppled over 15 million trees in southeast England alone and left a swathe of destruction in northern France, Belgium, and the Netherlands. Eighty percent of low-lying Bangladesh was inundated by floodwater in 1988. Tens of thousands of people drowned in a tropical cyclone that hit the same area in 1991. Hurricane Andrew devastated Florida in 1992 and caused $16 billion in damage. The Mississippi and Missouri Rivers rose to record levels in 1993 and flooded an area of agricultural land as large as one of the Great Lakes. El Niño–caused droughts wreaked havoc in tropical Africa and Southeast Asia, where huge forest fires raged out of control in 1982–1983 and 1997–1998.

Prophets of ecological doom are now out in force, while scientists passionately debate the extent of global warming and its effects on long-term climate. At a major conference in Rio de Janeiro, Brazil, in 1992, 154 nations agreed on an important convention that called for a stabilization of greenhouse gas concentrations in the atmosphere to prevent human inference with global climate. The Rio de Janeiro conference and a 1997 meeting in Kyoto, Japan, have generated vigorous controversy among nations and intense lobbying by special interests, especially large global corporations involved with oil production. Global warming now appears as a topic on political agendas. Vice President Al Gore of the United States recently proclaimed: "We know that as a result of global warming, there is more heat in the climate system, and it is heat that drives El Niño." He added confidently (and possibly inaccurately): "Unless we act we can expect more extreme weather in the years ahead."[3] The vice president was drumming up support for passage of a $6.3 billion program to reduce emissions of global warming gases and can be excused his dire prediction. One of the benefits of the recent El Niños is that they have brought the issue of global warming into the public spotlight.

Many people believe the fate of our civilization is closely tied to our ability to curb global warming. All of our climate models tell us that warming is inevitable if the concentration of greenhouse gases continues to increase at its present rate. Antarctic ice cores have trapped air bubbles that date back long before the Industrial Revolution. These bubbles show that carbon dioxide levels have risen sharply since 1850. Other greenhouse gases have increased at the same time, including methane, generated by the rise in human populations and dramatic increases in rice paddy agriculture and cattle herding.

Some people see the dramatic warming of the past two decades as a sign that anthropogenic global warming is already affecting our weather. Others take a longer view and point out that the present warm temperatures are no more extreme than those of the Medieval Warm Period of 800 years ago. They also argue that we have only about 150 years of accurate weather records to work with, all of them taken since the Industrial Revolution began. These figures show increasing temperatures between 1900 and the 1940s, then a slight dip (which had many people talking of an imminent Ice Age), followed by the rising temperatures of today. Are these temperature variations the result of human activity or simply natural climatic variability, like that documented by corals and tree rings over the past ten thousand years? We will not know until the warming trend reaches such levels that the evidence of human intervention is unequivocal. George Philander uses theoretical models of the concentrations of atmospheric gases to make three predictions:

1. Global average surface temperatures will rise 0.15 to 2 degrees Celsius between 1990 and 2050. We will experience more rainfall, warmer winter temperatures in the Arctic, and sea level rises of between three and forty centimeters by 2050.
2. Increases in higher-latitude rainfall will lead to more northerly farming in Canada and Siberia, while conditions will be drier in the middle latitudes of the Northern Hemisphere.
3. Short-term climatic changes will be more dramatic in small regions as the frequency of droughts, ENSOs, floods, and hurricanes increases.

If Philander is correct, then the greatest threats to humanity from global warming in the short term come not from millennial climate changes but from regional climatic shifts, such as an increase in El Niños, which have immediate, and usually adverse, effects on human populations. To some degree, the same has been true for the past ten millennia, but the stakes grow ever higher as the world's population continues to rise and marginal environments absorb more and more people. The Organization for African Unity estimates that at least half a million people perished from drought alone in sub-Saharan

Africa between 1980 and 1990, many of them as a consequence of the 1982–1983 ENSO event. History tells us that El Niños have sometimes provided the knockout punch that topples states and great rulers. How infinitely greater these kinds of stresses are in our over-populated and polluted world! If the scientists are right and global warming gives birth to more intense and more frequent ENSO events, then the fate of entire nations could lie in the unrelenting punches of the Christmas Child.

Many questions remain. Are we witnessing a dry run of climate changes caused by global warming? Or was the Southern Oscillation just being its usual unpredictable self with its more frequent El Niños of recent years? What causes the protean swings of the ENSO pendulum so convincingly modeled by today's computer simulations? Why do these models always evolve in different ways? Is El Niño a purely tropical phenomenon, or do some still-unknown external forces move the atmosphere and ocean, that George Philander calls "partners in the dance"? Who leads, the agile atmosphere or the more ponderous ocean? Oceanographer Richard Fairbanks likens the entire process to an orchestra playing a symphony. But nobody has found the conductor.

Oliver Goldsmith wrote in *Sweet Auburn:*

> *Ill fares the land, to hastening ills a prey,*
> *where wealth accumulates and men decay.*

What was true in Goldsmith's time two centuries ago is doubly true today. Unlike Goldsmith, we live in an overcrowded, industrial world where the "hastening ills" of humanity achieve new levels of folly every year. Our global perspective on El Niños allows us to assess the true extent of damage wrought by the strong ENSOs of recent years, with the promise of more frequent episodes in the future as global warming intensifies. The material destruction does not pose the greatest danger to humanity. But in societies already stressed by unwise management of the environment, an El Niño can destroy the people's faith in the legitimacy of their leaders and in the foundations of their society. Sometimes the leaders are wise, their institutions are flexible, and the society adapts successfully to new circumstances, as the Anasazi did. On occasion, the rulers fall but their successors learn from their mistakes, as the Theban nomarchs did when they laid the foundations of Middle Kingdom Egypt. Sometimes the stresses are relieved by some extraordinary event or a remarkable innovation, like the Younger Dryas–caused changeover to agriculture at Abu Hureyra or the large-scale cultivation of the potato in Europe during later stages of the Little Ice Age. But sometimes the collapse is total, as it was with the Moche and the Maya.

The stresses are mounting dangerously in our own world. Like Stone Age foragers at the end of the Ice Age, we have just about run out of space to run away from our troubles. No one force—overpopulation, global warming, or rapid climate—will destroy our civilization. But the combination of all three makes us prey to the knockout blow that could.

So we have several choices. We can simply hope the knockout blow never comes. We can hope that our technological brilliance—especially in information technology and biotechnology—is the extraordinary innovation that provides an escape hatch. But we should always remember that the societies that sustained themselves over the true long haul—over millennia, as the Africans of the Sahel did before European colonization—did so by reaching an accommodation with their environment.

The uncertainties that confront us are daunting, but with the crisis not yet upon us, there is still hope. For thousands of years we have been predators on earth, taking and exploiting rather than giving back, using up finite resources on the mistaken assumption that we were the appointed masters of the world. We have destroyed forests; eroded topsoil; polluted and altered the atmosphere; poisoned oceans, rivers, and lakes; and let the worst effects of our despoliation fall on the poor. But crises can bring out remarkable qualities in rulers and their subjects. The development of our ability to forecast El Niños is one such response. Perhaps the now-widespread public consciousness of global warming and of El Niños will lead to unprecedented levels of cooperation and commitment among people and nations to create a self-sustaining world. The record of the past suggests that the fate of industrial civilization depends upon it.

Humans have adapted exquisitely to the global environment for the past ten thousand years, but at a high price. In the past two centuries the Industrial Revolution has trapped us, through no one's fault, on an apocalyptic path that threatens our very existence. To escape this trap will require extraordinary solutions that transcend politics, religion, and individual goals. Unfortunately, history has no precedents to guide us, beyond reinforcing the belief that *Homo sapiens sapiens* is capable of rising to the challenge. As Otto von Bismarck remarked more than a century ago, we need to hear the footsteps of history and learn from them.

Notes and
Sources

The research for this book involved personal interviews, massive e-mail correspondence, a huge literature in many disciplines, and a maze of archaeological and historical sources. A curious reader will have to track down a complex and scattered literary archive. At present, only Michael Glantz's admirable *Currents of Change: El Niño's Impact on Climate and Society* (Cambridge: Cambridge University Press, 1996) offers a general account of ENSO for a broad audience, although we will doubtless see many more books in coming years. As a general rule, and in the interests of space, I have quoted only one book if there are several on a broad topic like, say, the history of Greenland, on the presumption that the reader will soon identify competing volumes. For each chapter, I have cited quoted passages and provided a summary of important sources that will lead an interested reader to more specialized references. The World Wide Web is a notable source of information on El Niño and other climatic phenomena, and the reader may use any of the common search engines to locate the latest presentations. Addresses given here would be too transitory to be meaningful.

Chapter 1
The Great Visitation

Accounts of the catastrophic famines of 1896–1897 and 1899–1900 abound, especially in the missionary literature. Many of these books are surprisingly graphic in their descriptions, with harrowing photographs of the dead and dying carefully calculated to attract relief funds. Nevertheless, they contain information of considerable historical value. Notable titles include: George

Lambert, *India: The Horror-Stricken Empire* (Berne, Ind.: Mennonite Volk Concern, 1898) and F.H.S. Merewether, *A Tour Through the Famine Districts of India* (Philadelphia: J. B. Lippincott Co., 1898). J. E. Scott, *In Famine Land* (New York: Harper, 1904), is a more sympathetic portrayal of the Curzon administration's relief efforts. See also Vaughan Nash, *The Great Famine and Its Consequences* (London: Longman Green, 1900).

The more general literature on Indian famines is extensive, including B. M. Bhatia, *Famines in India, 1860–1943* (New York: India Publishing House, 1967), and H. S. Srivastava, *The History of Indian Famines, 1858–1918* (Agra, India: Sir Ram Mehra, 1968). Michele Burge McAlpin, *Subject to Famine: Food Crises and Economic Change in Western India, 1860–1920* (Princeton, N.J.: Princeton University Press, 1983), is a perceptive analysis of changing British strategies to cope with monsoon famines. Premansukumar Bandyopadhyay, *Indian Famine and Agrarian Problems* (Calcutta: Star Publications, 1987), contains a critique of the British administration and numerous telling statistics. Richard Grove, "Global Impact of the 1789–1793 El Niño," *Nature* 393 (1998): 318–319, discusses the wide devastation caused by the late-eighteenth-century droughts.

Jay S. Fein and Pamela L. Stephens, eds., *Monsoons* (New York: John Wiley Interscience, 1987), provides an invaluable overview of monsoons and research into them. In particular, Khushwant Singh, "The Indian Monsoon in Literature" (35–50), was an invaluable guide. Historical perspectives come from two other chapters of this volume: Bruce Warren, "Ancient and Medieval Records of the Monsoon Winds and Currents of the Indian Ocean" (135–158), and Giela Kutzbach, "Concepts of Monsoon Physics in Historical Perspective: The Indian Monsoon (Seventeenth to Early Twentieth Century)" (159–210).

For a general history of India, try Stanley Wolpert, *A New History of India,* 5th ed. (New York: Oxford University Press, 1997).

1. Rudyard Kipling, *Rudyard Kipling's Verses* (Garden City, N.Y.: Doubleday/Doran, 1934), 90.

2. Quoted in Khushwant Singh, "The Indian Monsoon in Literature," in Jay S. Fein and Pamela L. Stephens, eds., *Monsoons* (New York: John Wiley, 1987), 45.

3. W. G. Archer, ed., *The Love Songs of Vidyapati* (London: Allen and Unwin, 1963), 100.

4. Quoted in Singh, "The Indian Monsoon in Literature," 47–48.

5. E. M. Forster, *The Hill of Devi* (London: Edward Arnold, 1953), 111. My father published Forster's novels and knew them almost by heart.

6. Both quoted in J. E. Scott, *In Famine Land* (New York: Harper Brothers, 1904), 2, 3.

7. Vaughan Nash, *The Great Famine and Its Consequences* (London: Longman Green, 1900), 62.

8. Both quoted in Scott, *In Famine Land,* 4, 11.

9. Ibid., 111.

10. Ibid., 127.

11. F.H.S. Merewether, *A Tour Through the Famine Districts of India* (Philadelphia: J. B. Lippincott, 1898), 135.

12. Scott, *In Famine Land,* 145.

13. Quoted in Bruce Warren, "Ancient and Medieval Records of the Monsoon Winds and Currents of the Indian Ocean," in Fein and Stephens, *Monsoons,* 144.

14. Quoted in ibid., 145.

15. Edmund Halley, "Historical Account of the Trade Winds and Monsoons," *Philosophical Transactions of the Royal Society* 16 (1686): 165.

16. Alexander von Humboldt, "Récherches sur les causes des inflexions de lignes isotherms," *Mémoirs de Physique de la Societé d'Arceuil* 3 (1817): 111.

17. Quoted in Ralph Abercromby, *Weather: A Popular Exposition of the Nature of Weather Changes from Day to Day* (London: Kegan Paul, 1887), 234.

18. Norman Lockyer, "Simultaneous Solar and Terrestrial Changes," *Native* 69 (1904): 351.

19. Henry Blanford, "On the Connexion of the Himalaya Snowfall with Dry Winds and Seasons of Rainfall in India," *Proceedings of the Royal Society of London* 37 (1884): 20.

20. Sir Gilbert Walker, "Correlations in Seasonal Variations of Weather, IX," *India Meteorological Service Memoirs* 24, no. 4 (1923): 22.

21. Ibid., 22.

22. Sir Gilbert Walker, "Correlations in Seasonal Variations of Weather, IX," *India Meteorological Service Memoirs* 24, no. 9 (1924): 19.

23. Walker's formulas were based on the notion that pressure oscillations return to a previous state, as opposed to statistical calculations based on probability theory.

24. Sir Charles Normand, "Monsoon Seasonal Forecasting," *Quarterly Journal of the Royal Meteorological Society* 79 (1953): 469.

25. Quoted in B. G. Brown and R. W. Katz, "The Use of Statistical Methods in the Search for Teleconnections: Past, Present, and Future," in Michael H. Glantz, R. W. Katz, and N. Nicholls, eds., *Teleconnections Linking Worldwide Climate Anomalies* (Cambridge: Cambridge University Press, 1991), 371–400.

Chapter 2
Guano Happens

The guano trade is a fascinating sideline of nineteenth-century history, much neglected by historians. In writing this short account, I drew on Jimmy M. Skaggs, *The Great Guano Rush* (New York: St. Martin's Griffin, 1994), which has a comprehensive bibliography. R. C. Murphy, *Bird Islands of Peru: The Record of a Sojourn on the West Coast* (New York: G. P. Putnam Sons, 1925), was also informative, as is the same author's reprinted "The Guano and Anchoveta Fishery," in Michael H. Glantz and J. Dana Thompson, eds., *Resource Management and Environmental Uncertainty* (New York: John Wiley Interscience, 1981), 81–106. This volume also contains other valuable articles on upwelling and fisheries. Edward A. Laws, *El Niño and the Peruvian Anchovy Fishery* (Sausalito, Calif.: University Science Books, 1997), is a simple guide to the subject.

General works on El Niño include Michael Glantz's valuable and widely read *Currents of Change: El Niño's Impact on Climate and Society* (Cambridge: Cambridge University Press, 1996), and George Philander's more technical *El Niño, La Niña, and the Southern Oscillation* (New York: Academic Press, 1990). Philander's *Is the Temperature Rising?* (Princeton, N.J.: Princeton University Press, 1998), is an eloquent discussion of global warming and world climate for laypeople that places El Niño in a wider context.

1. *Guano* is a corruption of the Spanish word *huano,* itself of Native American derivation, meaning "excrement of sea fowl."

2. R. A. Murphy, "Oceanic and Climatic Phenomena Along the West Coast of South America During 1925," *Geographical Review* 46 (1926): 26–54.

3. Camilo Carrillo, "Disertacíon sobre las Corrientes Océanicas y Estudios de Corriente Peruana de Humboldt," *Boletinas del Sociedad Geographico Lima* 11 (1892): 84.

4. Frederico Alfonso Pezet, "The Countercurrent 'El Niño' on the Coast of Northern Peru," *Boletinas del Sociedad Geográphico Lima* 11 (1895): 603.

5. Murphy, "Oceanic and Climatic Phenomena," 50.

6. Ephraim Squier, *Travels in Peru* (New York: Harper, 1888), 110, 129.

7. The conquistador Francisco Pizarro and his motley band of adventurers fed their horses off blooming desert forage as they rode inland to the heart of the Inca empire in 1532. Scientists debate whether this was an El Niño year. Perhaps the course of history would have been different if the desert had been its usual arid self.

8. Sir Gilbert Walker, "Correlations in Seasonal Variations of Weather, IX," *India Meteorological Department Memoirs* 24, no. 9 (1924): 3.

Chapter 3
ENSO

See Chapter 2 sources, especially Glantz, *Currents of Change,* and Philander, *Is the Temperature Rising?*

1. Quoted in *Latitude* 38 (October 1997): 132.

2. George S. Philander, *Is the Temperature Rising?* (Princeton, N.J.: Princeton University Press, 1998), 108.

Chapter 4
The North Atlantic Oscillation

The North Atlantic Oscillation has generated an enormous, highly technical literature. A series of articles by Wallace S. Broecker are invaluable. "Chaotic Climate," *Scientific American* (November 1995): 62–68, surveys the Great Ocean Conveyor Belt; "What Drives Climatic Cycles?" *Scientific American* (January 1990): 49–56, discusses links between the ocean, the atmosphere, and Ice Age climatic change. Also invaluable is Oliver Morton, "The Storm in the Machine," *New Scientist* 2119 (1998): 22–27, which summarizes much new research. Philander, *Is the Temperature Rising?,* offers a useful summary.

Michael H. Glantz, R. W. Katz, and N. Nicholls, eds., *Teleconnections Linking Worldwide Climatic Anomalies* (Cambridge: Cambridge University Press, 1991), is an excellent source.

1. Subsequent research has dated the Leman and Ower point to about 8500 B.C.

2. Finn Gad, *The History of Greenland,* vol. 1, *Earliest Times to 1700* (London: C. Hurst, 1970), 366.

3. Osbert Lancaster, *Daily Express,* December 17, 1947.

4. British Admiralty, *Ocean Passages for the World* (Taunton, Eng.: H. M. Hydrographic Office, 1950), 151.

5. Robert H. Fuson, *The Log of Christopher Columbus* (Camden, Maine: International Marine Publishing Co., 1987), 63.

Chapter 5
A Time of Warming

The Kalahari San are described in an enormous literature; Richard Lee, *The !Kung San* (Cambridge: Cambridge University Press, 1971), is a classic source.

Hxaro is analyzed by Polly Weissner, "Risk, Reciprocity, and Social Influences on !Kung San Economics," in Eleanor Leacock and Richard Lee, eds., *Politics and History in Band Societies* (Cambridge: Cambridge University Press, 1981), 61–85. For some general background, see George Silberbauer, "Neither Are Your Ways My Ways," in Susan Kent, ed., *Cultural Diversity Among Twentieth-Century Foragers* (Cambridge: Cambridge University Press, 1996), 21–64.

The Younger Dryas is described in many specialist papers. Wallace S. Broecker and George H. Denton, "What Drives Glacial Cycles?" *Scientific American* (January 1990): 49–56, offers an excellent account for the general reader. The Abu Hureyra monograph by Andrew Moore, G. C. Hillman, and A. J. Legge, *Abu Hureyra and the Advent of Agriculture* (New York: Oxford University Press, 1998), is seminal. A general account of the site will be found in my *Time Detectives* (New York: Simon & Schuster, 1995).

William Ryan and Walter Pitman, *Noah's Flood: The New Scientific Discoveries About the Event That Changed History* (New York: Simon and Schuster, 1998), describes the extraordinary detective work that went into the reconstruction of the Black Sea catastrophe of 5500 B.C. For a general account of early agriculture in Europe, see Alisdair Whittle, *Europe in the Neolithic: The Creation of New Worlds* (Cambridge: Cambridge University Press, 1996). A general account of early agriculture in the Old World can be found in my college textbook *People of the Earth,* 9th ed. (New York: Addison Wesley Longman, 1997). Samuel Kramer, *The Sumerians* (Chicago: University of Chicago Press, 1963), is still the most readable account of Sumer for the layperson, even if it is somewhat outdated. The controversial subject of the antiquity of El Niños is discussed by T. J. DeVries and others in "Determining the Early History of El Niño," *Science* 276 (1996): 295–297, where a debate will be found, as well as a full bibliography of more specialized references.

1. The exclamation mark denotes a palatal click.

2. This rate would increase to 0.1 percent after farming began in 9000 B.C., and to 0.6 percent and 2.0 percent in the nineteenth and twentieth centuries A.D., respectively.

3. Unfortunately, Abu Hureyra was flooded by a lake formed by a dam built in the late 1970s across the Euphrates. Fortunately, the site was brilliantly excavated by the archaeologist Andrew Moore in the 1970s before it vanished under water.

4. Computer simulations show that the transition from wild to domesticated grain may have occurred within a mere twenty or thirty years. Archaeologists will never find the first farming village of all, because the same transition probably occurred at many locations more or less simultaneously.

5. William Ryan and Walter Pitman have recently argued that the Black Sea catastrophe is a folk memory of the biblical great flood (*Noah's Flood: The New Scientific Discoveries About the Event That Changed History* [New York: Simon & Schuster, 1998]). I think most archaeologists would call this an intellectual stretch.

6. Samuel Kramer, *The Sumerians* (Chicago: University of Chicago Press, 1963), 71.

Chapter 6
Pharaohs in Crisis

Karl W. Butzer, *Early Hydraulic Civilization in Egypt* (Chicago: University of Chicago Press, 1976), was an essential source for this chapter. Fekri A. Hassan, "Nile Floods and Political Disorder in Early Egypt," in H. Nüzhet Dalfes, George Kukla, and Harvey Weiss, eds., *Third Millennium B.C. Climate Change and Old World Collapse* (Berlin: Springer-Verlag, 1994), 1–24, is a useful summary of climate and Old Kingdom Egypt. Butzer's "Sociopolitical Discontinuity in the Near East c. 2200 B.C.E.: Scenarios from Palestine and Egypt," in Dalfes, Kukla, and Weiss, *Third Millennium B.C. Climate Change* (245–296), argues for a more complex scenario in which long-distance trade and political and social change also played a part. Barbara Bell's two closely argued articles on climate and Egyptian history are seminal: "The Dark Ages in Ancient History I. The First Dark Age in Egypt," *American Journal of Archaeology* 75 (1971): 1–26; and "Climate and the History of Egypt: The Middle Kingdom," *American Journal of Archaeology* 79 (1975): 223–269. Sir William Willcocks, *Sixty Years in the East* (London: Blackwood, 1935), contains interesting accounts of Egyptian irrigation before the building of the first Aswan Dam. Mark Lehner, *The Complete Pyramids* (London: Thames and Hudson, 1997), is definitive. Barry Kemp, *Ancient Egypt: The Anatomy of a Civilization* (London: Routledge, 1989), is a superb analysis of Ancient Egypt that is especially good on the Old and Middle Kingdoms.

1. Unless otherwise stated, all quotes in this chapter are from Barbara Bell, "The Dark Ages in Ancient History I. The First Dark Age in Egypt," *American Journal of Archaeology* 75 (1971): 1–26, where full citations will be found.

2. The sandbank of Apophis was a place in the underworld where the dragon-serpent Apophis threatened to devour the sun god Ra each night.

3. Sir William Willcocks, *Sixty Years in the East* (London: Blackwood, 1935), 135.

4. For obvious reasons, this chapter describes the Nile Valley before the building of successive dams at Aswan in Upper Egypt in the twentieth century altered flood patterns and irrigation.

5. Giovanni Belzoni, *Narrative of the Operations and Recent Discoveries Within the Pyramids, Temples, Tombs, and Excavations in Egypt and Nubia* (London: John Murray, 1821), 301.

6. *The Imperial Gazetteer of India 1908: The Indian Empire* (Oxford: Clarendon Press, 1908), 1:127.

7. I.E.S. Edwards, *The Pyramids* (New York: Viking, 1985), 22.

8. J. A. Wilson, *The Culture of Ancient Egypt* (Chicago: University of Chicago Press, 1957), 126.

9. Egyptologists conventionally divide Ancient Egyptian civilization into four broad eras, separated by three periods of political confusion: (1) Archaic and Old Kingdom 3000–2134 B.C., First Intermediate Period 2134–2040 B.C.; (2) Middle Kingdom 2040–1640 B.C., Second Intermediate Period 1640–1530 B.C.; (3) New Kingdom 1530–1070 B.C., Third Intermediate Period 1069–525 B.C.; (4) Late Period 525–332 B.C.

10. Unless, as some believe, the scribe recording details of Pepi's reign confused the numbers 64 and 94, which are very similar in cursive hieroglyphs! Even 64 years is a long period of rule by any standards. Adolf Erman, *The Ancient Egyptians: A Sourcebook of Their Writings,* translated by A. M. Blackman (New York: Harper Torchbooks, 1927), 110.

11. Quoted in Peter Clayton, *Chronicle of the Pharaohs* (London: Thames and Hudson, 1994), 84.

12. Quoted in Barbara Bell, "Climate and the History of Egypt: The Middle Kingdom," *American Journal of Archaeology* 79 (1975): 261.

Chapter 7
The Moche Lords

Michael Moseley, *The Incas and Their Ancestors* (London: Thames and Hudson, 1992), is the best general source on Andean archaeology. The Moche civilization is summarized in Walter Alva and Christopher Donnan, *The Royal Tombs of Sipán* (Los Angeles: Fowler Museum of Cultural History, 1993), a magnificent account of the remarkable warrior-priest burials. Donnan's *Moche Art and Iconography* (Los Angeles: Institute of Latin American Studies, UCLA, 1978), is an authoritative account of their ceramics, with much relevance to understanding their rulers. Izumi Shimada, *Pampa Grande and the Mochica Culture* (Austin: University of Texas Press, 1994), is another valuable synthesis of the subject. See also Adriana von Hagen and Craig Morris, *The Cities of the Ancient Andes* (London: Thames and Hudson, 1998). The Moche civilization and El Niño is the subject of a widely scattered literature. Fred Nials et al., "El Niño: The Catastrophic Flooding of Coastal Peru, Parts I and II," *Field Museum Bulletin* 50, no. 7 (July–August 1979): 4–14, 50; and no. 8 (September 1979): 4–10, are useful starting points. For ice cores, see L. Thompson et al., "El Niño–Southern Oscillation and Events Recorded in the Stratigraphy of the Tropical Quelccaya Ice Cap," *Science* 226 (1984): 50–53, and the same authors' "A 1,500-Year Tropical Ice Core Record of Climate: Potential Relations to Man in the Andes," *Science* 234 (1986): 361–364. See also Michael E. Moseley, "Punctuated Equilibrium: Searching the Ancient Record for El Niño," *Quarterly Review of Archaeology* 8 (1987): 7–10. Izumi Shimada et al., "Cultural Impacts of Severe Droughts in the Prehistoric Andes: Application of a 1,500-Year Ice Core Precipitation Record," *World Archaeology* 22, no. 3 (1991): 247–265, is invaluable.

Chapter 8
The Classic Maya Collapse

The Maya are among the most accessible of early civilizations for the general reader, thanks to a plethora of good popular works. Michael Coe's *The Maya,* 4th ed. (London: Thames and Hudson, 1993), is a standard work, while the same author's *Mexico,* 3rd ed. (London: Thames and Hudson, 1994), covers a wider canvas. Linda Schele and David Freidel, *A Forest of Kings* (New York: William Morrow, 1990) is a classic popular account of Maya civilization using both glyphs and archaeology. Scott Fedick, ed., *The Managed Mosaic* (Salt Lake City: University of Utah Press, 1996), contains valuable essays on Maya agriculture, notably Fedick's introduction and conclusion to the volume. K. Anne Pyburn, "The Political Economy of Ancient Maya Land Use" (236–249), is especially useful. Kent V. Flannery, ed., *Maya Subsistence* (New York: Academic Press, 1982), with its valuable case studies, amplifies the Fedick volume in many important ways.

The evidence for drought in the Maya lowlands is surveyed by David A. Hodell, Jason H. Curtis, and Mark Brenner, "Possible Role of Climate in the Collapse of Classic Maya Civilization," *Nature* 375 (1995): 391–347, where a bibliography of more specialized references will be found.

1. Sylvanus Morley, *The Ancient Maya* (Stanford, Calif.: Stanford University Press, 1946), 141.

2. Dennis Tedlock, ed., *Popol Vuh* (New York: Simon & Schuster, 1996), 64.

3. Maya civilization continued to flourish in the northern Yucatán until the Spanish Conquest of the sixteenth century.

Chapter 9
The Ancient Ones

Tessie Naranjo, "Thoughts on Migration by Santa Clara Pueblo," *Journal of Anthropological Archaeology* 14 (1995): 247–250, a short essay on the notion of movement, has had a profound effect on archaeological thinking about the Anasazi. In the same issue, see also Catherine M. Cameron, "Migration and the Movement of Southwestern Peoples" (104–124). This entire number of the *Journal of Anthropological Archaeology* contains valuable essays on Anasazi population movements, from both Chaco and the Mesa Verde region. Linda Cordell, *Archaeology of the Southwest,* 2nd ed. (San Diego: Academic Press, 1997), is a fundamental summary, as is Stephen Plog's *Ancient Peoples of the American Southwest* (London: Thames and Hudson, 1997). For a review of Chaco road research, see Gwinn Vivian, "Chacoan Roads: Morphology," and "Chaco Roads: Function," *The Kiva* 63, no. 1 (1997): 7–34, 35–67.

Climatic data are well summarized by Jeffrey S. Dean, "Demography, Environment, and Subsistence Stress," in Joseph A. Tainter and Bonnie Bagley Tainter, eds., *Evolving Complexity and Environmental Risk in the Prehistoric Southwest* (Reading, Mass.: Addison Wesley, 1996), 25–56. Dean puts forward the model of long- and short-term climatic change in "A Model of Anasazi Behavioral Adaptation," in George Gumerman, ed., *The Anasazi in a Changing Environment* (Cambridge: Cambridge University Press, 1988), 25–44. This entire volume contains valuable chapters, including Dean's "Dendrochronology and Climatic Reconstruction on the Colorado Plateaus" (119–167). See also Jeffrey S. Dean and Garey S. Funkhauser, "Dendroclimatic Reconstructions for the Southern Colorado Plateau," in W. J. Waugh, ed., *Climate Change in the Four Corners and Adjacent Regions* (Grand Junction, Colo.: Mesa State College, 1994), 85–104.

1. Both quotes from Tessie Naranjo, "Thoughts on Migration by Santa Clara Pueblo," *Journal of Anthropological Archaeology* 14 (1995): 249–250.

2. Gustav Nordenskiöld, *The Cliff Dwellers of the Mesa Verde, Southwestern Colorado* (New York: AMS Press, 1973), 33.

3. Naranjo, "Thoughts on Migration by Santa Clara Pueblo," 250.

Chapter 10
The Little Ice Age

John Kingston, *The Weather of the 1780s over Europe* (Cambridge: Cambridge University Press, 1988), is a marvelous piece of meteorological detective work, complete with synoptic weather maps. Jean M. Grove, *The Little Ice Age* (London: Routledge, 1988), is an authoritative source on the subject with comprehensive references. Emmanuel Le Roy Ladurie, *Times of Feast, Times of Famine: A History of Climate Since the Year 1000* (Garden City, N.Y.: Doubleday, 1971), is a classic work that treats of the Little Ice Age and its impact on Europeans. Climatologically, it is somewhat dated, but the historiography is authoritative and useful. Fernand Braudel, *Civilization and Capitalism, Fifteenth–Eighteenth Century*, vol. 1, *The Structures of Everyday Life* (London: Collins, 1981), is a mine of information on European diet, famine, and economic life. John D. Post, *The Last Great Subsistence Crisis in the Western World* (Baltimore: Johns Hopkins University Press, 1977), provides authoritative background. On Greenland and the Norse, see Finn Gad, *The History of Greenland*, vol. 1, *Earliest Times to 1700* (London: C. Hurst, 1973). Hermann Flohn and Roberto Fanteshi, eds., *The Climate of Europe: Past, Present, and Future* (Dordrecht: D. Redel, 1984), discusses European climate over the past one thousand years. S. L. Swan, "México in the Little Ice Age," *Journal of Interdisciplinary History* 11 (1981): 633–648, offers a perspective on another area of the world.

1. A somewhat misleading, but commonly used, term coined by the glaciologist F. E. Matthes in 1942.

2. Magnus Magnusson and Hermann Pálsson, *The Vinland Sagas* (Baltimore: Pelican Books, 1965), 50.

3. Quoted in A.E.J. Ogilvie, "Climate and Economy in Eighteenth-Century Iceland," in C. Delano Smith and M. Parry, eds., *Consequences of Climate Change* (Nottingham, Eng.: Department of Geography, Nottingham University, 1981), 59.

4. Fernand Braudel, *Civilization and Capitalism, Fifteenth–Eighteenth Century*, vol. 1, *The Structures of Everyday Life* (London: Collins, 1981), 110.

5. Ibid., 117.

6. All quotations in this section are from Ladurie, *Times of Feast, Times of Famine*, 171ff.

7. Guy de la Bédoyère, ed., *The Diary of John Evelyn* (Woodbridge, Eng.: Boydell Press, 1995), 267.

8. Sir John Sinclair, *Statistical Account of Scotland* (Edinburgh, 1791–1799), 231.

9. The French *Encyclopédie* of 1765 proclaimed that potatoes were poor people's food, despite causing flatulence: "But what is a little wind to the vigorous organs of the peasants and workers!"; Braudel, *Civilization and Capitalism,* 1:170.

Chapter 11
"Drought Follows the Plow"

The literature on the Sahel is enormous, frequently repetitive, and sometimes self-serving, so discerning even a partial consensus is a challenge. John Reader's *Africa: A Biography of the Continent* (New York: Alfred A. Knopf, 1998), is quite simply the best general book on Africa ever written and a wonderful starting point. Michael H. Glantz, ed., *Drought Follows the Plow* (Cambridge: Cambridge University Press, 1994), contains valuable essays on desertification and marginal lands around the world. The same author's edited *Desertification: Environmental Degradation in and Around Arid Lands* (Boulder, Colo.: Westview Press, 1977), focuses mainly on Africa. The National Research Council's *Environmental Change in the West African Sahel* (Washington, D.C., 1984), is an admirable summary for a wide scientific audience. Michael H. Glantz, ed., *The Politics of Natural Disaster* (New York: Praeger, 1974), covers the 1972 famine and its political ramifications, with essays on everything from climatic change to agricultural production and medical care. See also William A. Dando, *The Geography of Famine* (New York: V. H. Winton, 1980), and Ronaldo V. Garcia, *Drought and Man: The 1972 Case History* (Oxford: Pergamon Press, 1981). Robert W. Kates, *Drought Impact in the Sahelian-Sudanic Zone of West Africa: A Comparative Analysis of 1910–1915 and 1968–1974* (Worcester, Mass.: Clark University, 1981), is a much-quoted study. James L.A. Webb Jr., *Desert Frontier: Ecological Change Along the Western Sahel, 1600–1850* (Madison: University of Wisconsin Press, 1995), contains valuable historical background. Earl Scott, ed., *Life Before the Drought* (Boston: Allen and Unwin, 1984), is useful in examining the impact of colonial governance. J. D. Fage, ed., *The Cambridge History of Africa,* vol. 2, *C. 500 B.C.–A.D. 1050* (Cambridge: Cambridge University Press, 1978), is a widely available historical work. See also Roland Oliver, *The African Experience* (London: Weidenfeld and Nicholson, 1991).

1. Rinderpest is a contagious viral disease that is almost invariably fatal to cattle. The disease existed in classical times and ravaged Europe and Southwest Asia periodically from a reservoir of infection on the central Asian

steppes. The Italian army imported infected cattle from India, Aden, and south Russia into Ethiopia in 1889 to feed their soldiers at Massawa. A fast-moving and virulent rinderpest epidemic swept across the Sahel and the rest of tropical Africa, killing millions of cattle, sheep, and goats, as well as many game populations. Rinderpest wiped out much of Africa's wealth on the hoof almost overnight. Many Fulani wandered distraught, calling imaginary cattle, or committed suicide. In East Africa, the Maasai lost two-thirds of their cattle. Many herders turned to agriculture for the first time but were too weak from hunger to plant and harvest crops.

2. Robert W. Kates, *Drought Impact in the Sahelian-Sudanic Zone of West Africa: A Comparative Analysis of 1910–1915 and 1968–1974* (Worcester, Mass.: Clark University, 1981), 110.

3. Ibid.

4. Randall Baker, "Information, Technology Transfer, and Nomadic Pastoral Societies," quoted in Michael Glantz, "Nine Fallacies of Natural Disaster: The Case of the Sahel," in Michael H. Glantz, ed., *The Politics of Natural Disaster* (New York: Praeger, 1974), 3–24.

5. Michael H. Glantz, ed., *Drought Follows the Plow* (Cambridge: Cambridge University Press, 1994).

Chapter 12
El Niños That Shook the World

The literature on both the science and the social impacts of the ENSOs of recent years continues to proliferate at breakneck speed. No one can possibly keep up with all of it. Here are some useful summaries that were of assistance in the writing of this chapter. Glantz, *Currents of Change*, contains a survey of the 1982–1983 ENSO and its impacts. Michael Glantz, Richard Katz, and Maria Krenz, eds., *The Societal Impacts Associated with the 1982–1983 Worldwide Climatic Anomalies* (Boulder, Colo.: National Center for Atmospheric Research, 1987), is a series of essays that describe 1982–1983 in different parts of the world. P. W. Glynn, ed., *Global Consequences of the 1982–1983 El Niño–Southern Oscillation* (Amsterdam: Elsevier, 1990), discusses oceanographic perspectives. The 1997–1998 ENSO will doubtless produce a flood of impact reports and books in the coming years, but at present the main sources are newspaper accounts and the World Wide Web. The *Los Angeles Times* and the *New York Times* have run many human interest stories that go beyond the all-too-common superficial accounts of death and destruction. Much of the research for this chapter came from a constant monitoring of the Web, including IWO and FEWS bulletins.

1. Jean-Paul Malingreau, "The 1982–1983 Drought in Indonesia: Assessment and Monitoring," in Michael Glantz, Richard Katz, and Maria Krenz, eds., *The Societal Impacts Associated with the 1982–1983 Worldwide Climatic Anomalies* (Boulder, Colo.: National Center for Atmospheric Research, 1987), 17.

2. Gilbert Walker, "Seasonal Weather and Its Prediction," *Smithsonian Institution Annual Report* (Washington, D.C.: Smithsonian Institution, 1935), 117.

3. Quoted in Famine Early Warning System (FEWS) bulletins (February 1998).

4. Quoted in Famine Early Warning System (FEWS) bulletins (June 1997).

Chapter 13
The Fate of Civilizations

While much of this chapter brings together ideas developed during the writing of this book, I also delved into the often polemical and frequently confusing literature on future climate change, anthropogenic global warming, and world food production. Two books stand out from the crowd and are excellent starting points for general readers. George Philander, *Is the Temperature Rising?*, is a model of sober, clear analysis, description, and forward thinking on climate change in the near and long-term future. John Houghton, *Global Warming: The Complete Briefing*, 2nd ed. (Cambridge: Cambridge University Press, 1997), is a comprehensive guide to what we really know about global warming and its impacts on human society. Houghton also addresses the thorny issues of what governments can do to mitigate the problem. Jared Diamond's *Guns, Germs, and Steel* (New York: W. W. Norton, 1997), is a provocative essay on human history that touches tangentially on many of the questions addressed here. Carl Sagan, *Billions and Billions* (New York: Random House, 1997), also discusses some of the issues in this chapter, as does Al Gore, *Earth in the Balance* (Boston: Houghton Mifflin, 1992). Crispin Tickell, *Climate Change and World Affairs*, 2nd ed. (Cambridge: Harvard University Press, 1986), is an important summary.

1. Jared Diamond, *Guns, Germs, and Steel* (New York: W. W. Norton, 1997), 408.

2. Large-scale irrigation projects were not a feature of early Sumerian and Egyptian civilization. They were instituted later, when populations had risen considerably.

3. Quoted in *U.S. News & World Report,* June 22, 1998, 58.

Index

	CLIMATIC EVENTS	HISTORICAL EVENTS
A.D. 2000 —	⊢ El Niños (1957 onward)	
1900 —	⊢ 1925 El Niño ⊢ 1899 Indian famine	Sahel famine (1914)
1800 —	1789-92 El Niño	Industrial Revolution French Revolution
1700 —	Little Ice Age	
1600 —	Major Indian monsoon failures	Subsistence crisis in Europe
1500 —		Norse abandon Greenland
1400 —		
1300 —	Great Drought in the Southwest	Mesa Verde Anasazi dispersal
1200 —	Drought cycle in the Southwest	Chaco Anasazi dispersal
1100 —	Medieval Warm Period	Chavin state affected by major El Niño
1000 —		Colonization of New Zealand Colonization of Greenland by the Norse
900 —	Drought cycle in Maya lowlands	
800 —		Classic Maya collapse in Mesoamerican lowlands
700 —		Moche state collapses
	Severe drought in Peru	
600 —	Severe drought in lowland Mesoamerica and Peru	